INSTRUCTION

OF

MUSKETRY.

ADJUTANT-GENERAL'S OFFICE, HORSE GUARDS,
1st January 1856.

The Naval & Military Press Ltd

Published by the
The Naval & Military Press
in association with the Royal Armouries

Unit 10 Ridgewood Industrial Park,
Uckfield, East Sussex, TN22 5QE
Tel: +44 (0) 1825 749494
Fax: +44 (0) 1825 765701

MILITARY HISTORY AT YOUR FINGERTIPS
www.naval-military-press.com
ONLINE GENEALOGY RESEARCH
www.military-genealogy.com
ONLINE MILITARY CARTOGRAPHY
www.militarymaproom.com

ROYAL ARMOURIES

The Library & Archives Department at the Royal Armouries Museum, Leeds, specialises in the history and development of armour and weapons from earliest times to the present day. Material relating to the development of artillery and modern fortifications is held at the Royal Armouries Museum, Fort Nelson.

For further information contact:
Royal Armouries Museum, Library, Armouries Drive,
Leeds, West Yorkshire LS10 1LT
Royal Armouries, Library, Fort Nelson, Down End Road, Fareham PO17 6AN

Or visit the Museum's website at
www.armouries.org.uk

In reprinting in facsimile from the original, any imperfections are inevitably reproduced and the quality may fall short of modern type and cartographic standards.

GENERAL ORDER.

HORSE GUARDS,
1st January 1856.

THE Field-Marshal Commanding in Chief desires that the accompanying revised edition of the " INSTRUCTION OF MUSKETRY" be circulated for general information and guidance.

General and other Officers in command of Districts and Stations will be pleased to make it a part of their duty to ascertain that the injunctions therein contained are strictly observed, both in letter and spirit, in all Regiments or Depôts armed with rifled muskets, and the result, as regards the proficiency of the troops in ball-firing, is to be recorded in their Confidential Reports.

By Command of The Right Honourable
FIELD-MARSHAL VISCOUNT HARDINGE,
Commanding in Chief,

G. A. WETHERALL,
Adjutant-General.

CONTENTS.

PART I.—DUTIES OF THE INSTRUCTORS OF MUSKETRY IN BATTALIONS.

	Page
The Officer-Instructor of Musketry	7
Assistant Officer-Instructor	8
Non-commissioned-Officer Instructor	8
The Company Instructor	9
Summary of Instruction	9

PART II.—THEORETICAL PRINCIPLES.

First Part	12
Second Part	19

PART III.—PRELIMINARY INSTRUCTION.

Drills.

Target Drill { Aiming Drill	27
{ Position Drill	28
Judging Distance Drill	30
Manufacture of Cartridges	33

Practice.

Target Practice	34
Individual firing	35
1st Period	39
2nd Period	40
3rd Period	41
File and Volley Firing	42
Skirmishing (Two Practices)	42
Remarks on Extent of Range	44
Judging Distance Practice	45
1st Period	48
2nd Period	48
3rd Period	48
Instruction of Recruits	49
Prizes	50
Returns	50
Recapitulation, showing the Number of Drills to be gone through annually, and the Amount of Ammunition to be expended	52
Appendix, showing the manner in which the Instruction is to be proceeded with	55
List of Articles sanctioned for Instructors	60

INSTRUCTION OF MUSKETRY.

PART I.

Duties of the Instructors of Musketry in Battalions.

1. IN every battalion the instruction in *firing*, placed like all other exercises under the responsibility of the commanding officer, will be especially confided to the direction of a captain, who, having been reported by the commandant of the school of musketry qualified to exercise the functions of *officer-instructor of musketry*, will be charged with the entire training in musketry of the young *officers* and *recruits*, and with the theoretical and preliminary instruction, annually, of the other officers and soldiers of the battalion. Officer-Instructor of Musketry.

2. The target practice of the soldiers, when practising by companies, will take place under the command of their captains; the officer-instructor, however, will be present to assist the captains by his superior knowledge in this particular branch, and in order that the instruction and practice may be conducted with uniformity throughout the battalion.

3. The commanding officer will assemble the officers of the battalion at least once in each half-year, and will cause the non-commissioned officers and men to be assembled occasionally by squads or companies, when the officer-instructor, having previously explained the theoretical principles contained in this book, will be at liberty to advance deeper into the subject, developing to a degree proportionate to the rank and intelligence of his auditors, the whole history of small arms from the first invention of gunpowder, and the successive steps by which the rifled musket has attained its present efficiency, in order that the officers and soldiers, by acquiring a thorough knowledge of the subject theoretically, may take greater interest in the practical part of this most important branch of their duty.

4. The officer-instructor will arrange the progression of the different companies from one branch of the instruction to another; he will cause the ground to be properly prepared for practice, and butts to be erected according to rule.

5. At every practice there will be a fatigue party on the ground for the management of the targets under the orders of the officer-instructor.

6. The officer-instructor and his assistants will inspect all the practice registers, diagrams, and returns, and see that they are correct and strictly according to form; he will also make out, for the commanding officer's information and signature, the battalion returns which are to be transmitted at the prescribed periods to the school of musketry, and submit for his approval the names of men eligible for the battalion and company rewards, or prizes for target and judging distance practice.

7. The officer-instructor will be exempted from all regimental duty, and from such garrison duty as may interfere with his duties as an instructor of musketry.

Assistant Officer-Instructor.
8. A subaltern in each battalion will be chosen to act as assistant instructor; he will be exempted from such garrison and regimental duties as may interfere with his duties as assistant instructor. This officer must have qualified himself for the appointment at the school of musketry.

Non-commissioned-Officer Instructor.
9. The non-commissioned officer selected for promotion to this rank will be sent to the school of musketry: should he be reported qualified for the position, he will be appointed non-commissioned-officer instructor of the battalion. He will be placed under the special orders of the officer-instructor, and will rank next to the quarter-master serjeant (in the Guards with the drill serjeants according to seniority). He will assist the officer-instructor in all his duties, attending all target drills and practices, and will superintend the company instructors in the performance of their duties. He will precede the companies to the practice ground with the fatigue party, and see that the targets, &c., are placed according to the orders received from the officer-instructor. He will measure the distances himself, and be responsible that they are correctly marked. He will take charge of the targets, measuring cords, flags, &c., so long as they remain in use, and until they are delivered over to the quarter-master's department, or to the barrack-master on the removal of the corps to another station.

10. If a diagram has been kept, the non-commissioned-officer instructor will compare it, first, with the target, and then with the register, and ascertain their correctness. He will receive, at the conclusion of each practice, the column "*duplicate total points*," of the registers; these he will make use of to check, at the conclusion of each period of target and judging distance practice, the companies' returns, which, if found correct, he will take to the officer-instructor, who, after careful inspection, will countersign them, and make out the battalion return from them. He will be exempted from all garrison and regimental duty.

11. The senior serjeant of each company will be charged with the instruction of the men in *target practice, judging distance,* and *cleaning arms,* under the orders of his captain and of the officer and non-commissioned-officer instructors of the battalion. He will keep the registers for his company on the practice ground, and at the conclusion of each practice will read over to the men the number of points obtained by each, after which he will compare his register, first with the target, and afterwards with the diagram in the marker's butt, should one have been kept; both register and diagram will then be signed by two serjeants, viz., the company-instructor and the serjeant who has marked in the butt; the company-instructor will then take them to the officer-instructor, who will countersign them. The Company Instructor.

12. The column "*duplicate total points*" of the register, after receiving the initials of the officer-instructor, will be immediately torn off and handed over to the non-commissioned-officer instructor of the battalion. The same form will be attended to in the practice of judging distance.

13. The company-instructor will attend with his company when it is ordered for target drill or judging distance drill, and assist the officer and non-commissioned-officer instructors in the performance of their duties.

Summary of Instruction.

14. The instruction of musketry is divided into two principal parts, the THEORETICAL and the PRACTICAL.

15. The theoretical branch is confided especially to the officer-instructor, who will explain the principles thereof in the manner pointed out under this head. In this instruction the soldier will be made to understand the

reasons for all those rules which have to be attended to in practice.

16. The practical branch is divided into two principal parts, DRILL and PRACTICE. In the former are comprised *cleaning arms, target drill, judging distance drill*, and the *manufacture of cartridges;* the latter is divided into *target practice* and *judging distance practice.*

17. Young officers and recruits are never to be permitted to fire until they have been instructed in the first eighteen paragraphs in Part II., and exercised in the first three drills here mentioned; nor are the soldiers of the battalion to be allowed to fire their annual allowance of practice ammunition until they have been first similarly exercised.

Cleaning Arms.
18. In this branch the soldier will be made to learn the names of the different parts of the firelock, and the rules for cleaning and keeping them in proper order, this will be the first exercise in which the soldier is practised; and the instruction therein will be given to the recruit by the non-commissioned-officer instructor of the battalion, and to the soldiers in the companies by the company instructors.

Target Drill.
19. Is divided into *aiming* and *position drills;* in the first, the soldier will acquire a knowledge of the use of the sights, and his progress in this branch will be tested by making him aim at different distances by means of a rest; in the second, the soldier will be put through all the motions of firing, standing and kneeling, with the same accuracy as if actually firing ball, attention being paid to each movement. This exercise is to habituate the soldier to the correct position, and to the natural connection that should exist between the hand and the eye; and is intended to make up in some measure for the small amount of target practice of which the yearly allowance of ball ammunition admits.

Judging Distance Drill.
20. In this drill the soldier will be accustomed to take note of the size and appearance of men and objects at different distances.

Manufacture of Cartridges.
21. In each company from ten to twelve men will be instructed in the manufacture of cartridges by the company instructor.

Target Practice.
22. Target practice is the *proof* of the attention that has been paid to the preliminary drills; it is divided into three parts, namely:— firing *singly, file firing and volleys,* and firing in *extended order* as skirmishers, in which, the

practice of judging distance is combined with that of target practice.

23. This practice has for its object, to test the proficiency of each company in judging distance, and, when possible, will be carried on, during target practice, by the sections that are not occupied in firing. *Judging Distance Practice.*

24. The details of all the above branches of instruction are laid down in subsequent parts of these regulations; they will each be gone through, yearly, by every soldier of the battalion, and the number of drills or parades to be devoted to each branch, with the amount of ammunition to be expended at each drill or parade, is given in a table at page 52. The time to be spent by the recruits in these exercises before they are dismissed drill must depend on the intelligence of the individual and the progress made, according to the discretion of the officer-instructor; but the amount of ammunition to be expended in the instruction of the recruits is not, under ordinary circumstances, to exceed the amount specified under the column for recruits in the table before alluded to.

25. To the above course of instruction may be added that of *skirmishing*, when the nature of the ground admits of it. A squad of from sixteen to twenty files will be extended in skirmishing order on broken ground and made to advance and retire; going through all the motions of firing, judging their distance, and setting the sights according to the position of a supposed enemy, or of another squad in front of them; the squad will occasionally be halted, when the instructor will go down the ranks to examine, and, if necessary, correct the position of each soldier, pointing out errors either in the appreciation of distances, or in the method of covering himself from the fire of the enemy by taking advantage of the irregularity of the ground; he will also require each man to point out to him the position he intends to occupy when next ordered to move, either in advancing or retiring.

26. After the target practice has been gone through, the men should (if there is any extra ammunition) be trained to use their muskets at short distances, from 50 to 300 or 400 yards, with the sights down, judging for themselves the proper elevation or depression to be given to their muskets.

PART II.

Theoretical Principles.

THE following course will be followed by officers in explaining the theoretical principles of musketry to their men:—

I.

1. A black board and a piece of chalk should be made use of to describe the figures.

Construction of the barrel.
2. The instructor will explain the construction of the barrel, so far as is necessary to show that its upper surface does not lie in the same direction as the inside or *bore*.

Axis of the piece.
3. That the *axis of the piece* is an imaginary line along the centre of the bore. It denotes the course taken by the centre of the bullet whilst under the influence of the exploded powder, and the *distance and direction in which it is guided in its flight*.

Line of fire.
4. Having drawn upon the board a section of the firelock, with the axis (A B, fig. 1,) laid horizontally, he will then draw the *line of fire* (B C) in continuation of the axis, and explain that this is the *direction in which the bullet would fly*, and with a uniform velocity, were it not *impeded* by the *resistance of the atmosphere* and *drawn from it* by the *force of gravity*.

Laws influencing the course of bullet.
5. He will explain that the *atmosphere* consists of a multitude of small particles which cannot be moved aside by the bullet without imparting to it some degree of reactionary force, and so *reducing at every moment the velocity of its flight*.

Resistance of the air.

Force of Gravity.
6. That the *force of gravity* commences to act upon the bullet as soon as it quits the muzzle, *drawing it towards the ground* with greater velocity in proportion as it becomes longer exposed to its influence.

7. That these two distinct motions, *the one increasing as the other diminishes*, cause the bullet to move in a *curved line*, called the *trajectory*. For a short distance, in consequence of the great velocity of the bullet in its forward motion, and the comparatively slight influence of gravitation, the course of the bullet scarcely deviates from the line of fire, but the curve increases more and more in proportion as it becomes distant from the muzzle.

Trajectory.

8. He will then draw the trajectory B D, upon the board.

9. Having now explained the course of the bullet in the air and the laws which influence it, the next step is to show how a knowledge of these laws can be applied in practice.

10. The instructor will make the men observe, that *if the axis of the piece is directed upon an object the bullet will never hit it*, but for reasons aforesaid will always pass below it; by experiment, it has been found that at 100 yards it will pass about 1 *foot 5 inches below it.* (Then draw C D, denoting the fall at 100 yards.) To hit a mark at 100 yards it will therefore be necessary that the line of fire should be directed to a point 1 *foot 5 inches above* that mark, in which case the trajectory, conforming to the movement of the line of fire, will also be raised 1 foot 5 inches, and the bullet will strike the object. (Here change the direction of the axis of the piece,* and draw the new line of fire to F, as directed; draw the new trajectory passing through the mark C. In order that the soldier may not be confused by too many lines, the old line of fire and the old trajectory, marked in dotted lines in fig. 1, may now be rubbed out; and the trajectory may be continued to any distance that may be necessary.) Initial direction to be given to a bullet to enable it to hit a mark.

11. *To aim with accuracy, it is necessary that the sights should be carefully aligned between the eye and the mark.* If the sights, however, on the upper surface of the barrel were so constructed as to be in the same direction, or in other words parallel to the axis, it would then be necessary at 100 yards to aim 1 foot 5 inches above the mark; but in so doing the firer would lose sight of the object, and he would besides be uncertain of his correct elevation. The height of the lowest back sight of the firelock is therefore so arranged, that when aim is taken straight upon an object at 100 yards the axis of the piece receives the necessary degree of elevation. (The instructor must then draw the line of sight, C, G, H, from the mark to the top of the fore sight and continue it to the back sight, the height of which he will arrange accordingly.) Arrangement of the sight for 100 yards, and line of sight.

12. Having marked a point or drawn a figure (K) upon the line of sight at any distance beyond the 100 yards point, and assigned to it a distance of 200 yards, he will make the men observe, that if the same sight were used for this distance as for 100 yards, and the aim taken straight Arrangement of the sights for different distances.

---

* N.B. The section of the firelock may be cut in cardboard and fastened to the board by a pivot at the breech, on which it can easily be turned in any required direction.

upon the mark, the bullet would pass below it, showing that the 100 yards sight does not give sufficient elevation for this increased distance. As at the distance of 100 yards, and indeed at all other distances, the line of fire must be directed as much above the object as the bullet passes below the line of fire at those distances, and in order that the soldier may be enabled to aim straight at the mark at all distances, the back sight is made capable of adjustment, so that it is thus available for all ranges from 100 to 900 yards; by which means, if the firer is certain of his distance, he need never trouble himself about his elevation, knowing that this is already arranged for him in the construction of the sights. Beyond 900 yards the soldier must judge for himself by raising his eye as much above the back sight as he may think necessary, always keeping the fore sight in line with the object. At distances under 100 yards (M), allowance may be made for the slight rise of the trajectory above the line of sight by aiming a few inches under the mark; at 50 yards about eight inches will be sufficient; at 30 yards about four inches.

13. In order to explain this more thoroughly to the soldier, the instructor will cause the breech of a barrel to be taken out and a plug of wood to be substituted in its place (fig. 2). In the centre of this he will bore a small hole for the soldier to look through, and at the muzzle of the gun a cap with cross wires should be fitted; he will make the soldier place the barrel in a traversing rest, and aim with the sights set to a given distance; then looking through the barrel he will observe the angle formed by the line of fire.

Necessity of holding the sights upright.

Error of direction.

14. Whilst the barrel is still upon the traversing rest the instructor will explain to the men the importance of holding the sights upright; having drawn a vertical line upon the target or board (fig. 3), he will tell the soldier to aim at a given spot upon this line, making use of the 900 yards sight, and taking care that it is *perfectly upright;* then looking through the barrel, he will observe that the cross wires *cut the vertical line above the mark.* The instructor will then shift the barrel and will cause the soldier to aim at the same mark and with the same sight, but *inclined to one side;* then looking again through the barrel, it will be seen that the line of fire, instead of being directed upon the same spot as before, is *directed to that side on which the sight is inclined,* and as the trajectory always conforms to the direction of the line of fire, the bullet, instead of

hitting the mark, will strike on that side. The greater the distance, the greater will be the error due to any inattention in this particular.

15. Not only is the *direction* affected by this inclination of the line of sight but the *elevation* also; this will be best explained by a card (fig. 4) with lines drawn upon it to represent the proper height of the back sight at each distance; placing this card vertically at the back of the sight, the instructor will make the men observe what amount of elevation is lost by any given degree of inclination of the back sight. Error of elevation.

16. Both the error of *direction* and of *elevation* may be shown by a small model fire-lock (fig. 5) with wires affixed to it, representing the line of fire, line of sight, and trajectory. If the trajectory is made to hinge upon the line of fire, it is obvious that it will show the exact course which the bullet will pursue at any degree of inclination of the back sight, *the trajectory always preserving the same relative position beneath the line of fire.*

17. The effect of wind should be next shown; when blowing from the right it will blow the bullet towards the left of the mark, and vice versâ, when from the front it will slightly reduce the speed of the bullet, and when from the rear increase it, causing in the first instance a *reduction* and in the latter an *increase* of the range; the soldier should, however, be cautioned, that a front or rear wind does not produce so much effect as a side wind, and that he must be guided entirely by his own experience in making allowance for wind, as no fixed rules can be laid down for his guidance. If the wind, for instance, is blowing from the left, he must aim a little to the left of the mark; if he finds that the shot still strikes to the right, he must make a little more allowance in firing the next shot. Error due to wind.

18. It has been found that the simplest way of illustrating the principles which regulate the adjustment and position of the sights is by means of a model gun, about a foot in length (fig. 10), having in the barrel a small spring projecting an iron bullet with sufficient force to penetrate a screen of silver paper at a short distance from the muzzle. A muzzle-stopper should be attached to it, having a ring in the centre, to which a string may be tied representing the line of fire. Another string should be used to show the line of sight. A folding back sight should be fitted on

the barrel, having three lines of sight—one on the surface of the barrel parallel to the axis, one arranged for the distance of 4 feet, and one for 8 feet. The gun should be first placed at the distance of 4 feet, and the line of fire directed upon a mark on the screen on the same level as the muzzle; remove the muzzle-stopper and fire the gun, when it will be seen that the bullet enters the paper below the mark. Measure the fall of the bullet and place a spot upon the screen as much above the object to be struck as the bullet has fallen below it, and direct the line of fire upon this spot, raising the muzzle; if the gun is now fired the object will be struck. The soldier should then be made to look along the sight which is parallel to the axis, and he will observe that the muzzle hides the object from view; he has therefore no security that the sight is properly directed, nor has he any guide to the elevation, except the previous measurement, which would, of course, be impracticable when firing with a musket. Now raise the folding sight, and raise the eye without altering the position of the gun until the foresight is in line with the mark; raise the brass sliding bar to the height of the eye, and make a mark on the sight showing the elevation due to the distance 4 feet, so that when it is again intended to fire at this distance the firer has only to raise the sliding bar to the mark on the sight and aim straight at the object. Let the gun now be removed to 8 feet distance, and make use of the same sight that has been previously adjusted for 4 feet, it will be seen that the bullet will strike low, showing the necessity of using a higher back sight for this distance, and the importance of knowing accurately the distance from the mark in order to regulate the sight correctly. The stand on which the gun is fixed should have a hinge at A, to enable the instructor to show the effect of inclining the sights to one side.

19. When the instructor has ascertained that all the men thoroughly understand the foregoing sections, and not until then, he may enter into an explanation of the various causes of uncertain firing in the old smooth bore musket, in order that he may show by what means these defects have been obviated by the introduction of rifled barrels.

Causes of uncertain firing with the old musket.

20. The chief cause of error was the excess of windage. By windage is meant the difference of size between the bore and the bullet; (here the instructor will draw a section of the inside of the barrel and the round bullet resting

upon its lower surface in contact with the powder (fig. 6). A certain amount of windage was necessary with the old musket, otherwise, when the barrel became foul after firing, the bullet would not enter. When the musket is held up to the "Present" the bullet will rest on the lower surface of the barrel, the whole of this windage consequently will be above. During the explosion of the powder, a portion of the gas will therefore escape by the windage forcing the bullet down upon the lower surface, at the same time that it forces it out of the barrel; the bullet, bounding from the lower surface of the barrel, will strike against the top, and so continue to rebound from the upper surface to the lower, or from side to side in a zig-zag direction, instead of passing evenly along the barrel; the result of this is, that, on whichever side the bullet last strikes the side of the barrel, it will experience a reactionary force tending to send it in an opposite direction, and so divert it from the course it is intended to pursue:— besides this a rotatory motion will be given to it, calculated to send it in the same direction. *Excess of windage.*

21. The bullet now used in the rifle musket (fig. 7), besides being elongated, and therefore better shaped for passing through the air, is so contrived that in its passage out of the barrel all windage is done away with. It enters the barrel easily, but on the explosion taking place, the pressure of the air in front, and the force of the explosion behind, have the effect of dilating the cylindrical portion of the bullet, so as to make it fit the barrel tightly, precisely in the same way that compressing an orange or India-rubber ball at the opposite ends would widen its other diameter and so enlarge its lateral circumference. *Error due to excess of windage rectified by the expanding bullet.*

22. By this means windage is completely done away with, and the whole force of the explosion acts upon the bullet in the same direction, giving it increased velocity, and avoiding any of those irregularities which have been described as taking place during the passage of the *spherical bullet* through the smooth bore barrel.

23. But besides these irregularities tending to give the spherical bullet a wrong direction, there is another cause calculated to influence it during its flight. Suppose a bullet (fig. 8) to be passing through the air in the direction of the arrow A, and that by some accident it had a hollow or unevenness on one side at B, this will receive the pressure of the atmosphere in the direction of the arrow *Error due to defective figure.*

B

C, which would tend to divert the bullet from its true course and send it in the direction of the arrow D. Now the present elongated bullet would be equally if not more affected by any such unevenness on its surface if it were fired out of a smooth barrel, but when fired out of a rifled barrel any error arising from this cause is corrected in the manner described in the next section.

The rifled barrel.

24. The rifled barrel is cut by three spiral grooves— (here the soldier should be made to look through the barrel and observe the grooves)—constructed in such a manner that the groove which is on the left side at the breech makes a half turn over the barrel, and appears on the right side of the muzzle; so the other two grooves make a half turn in the barrel, passing over like a female screw from left to right.

Error due to defective figure rectified by the rifled barrel.

25. When the bullet is expanded by the explosion of the powder, as before described, it is not only made to fit the barrel tightly, but its cylindrical surface is moulded into the grooves in such a way, that during its passage through the barrel it is constrained to turn with the grooves, and so it receives a spinning movement round its longer axis, which continues during the remainder of its flight; and this not only prevents any rotation in any other direction, but is in itself a rotation calculated to ensure accuracy of flight, by constantly presenting any imperfection of surface to the air in opposite directions.

26. The best way of describing the motions of a rifled projectile in its passage through the air is by means of a bent arrow, B (fig. 9). Suppose the arrow was shot from the point A, with a view to hit the mark C; if the bent point of the arrow were placed upwards, and no spinning motion given to it, the greater pressure of the air on its convex side would tend to send it in the direction D; but if it had previously received a spinning motion, by the time it reached E the point would have turned in an opposite direction, and therefore it would proceed downwards, crossing the true trajectory, and proceeding as far beneath it as it had at first risen above it; thus it would continue throughout its course to move in a spiral direction round the true trajectory, constantly correcting the error due to its imperfect figure, and ultimately it would strike the mark much nearer than if it had received no spinning motion.

27. The only object of *rifling* a barrel is to correct by this means the flight of the bullet; it does not, in itself, produce either greater range or velocity.

II.

28. In the foregoing part of the Theoretical Instruction, the soldier will have formed some idea of the laws which regulate the flight of the bullet in the air, and he will have been rendered thoroughly conversant with the arrangement of the sights and of the barrel, and complete master of all the rules which have been laid down for his guidance at the moment of firing.

29. The instructor must now explain to him, that the accuracy of his fire is not dependent on these causes alone, but also on the attention that he pays to the preservation of his arms and ammunition; besides which there are other sources of error inherent in the arm itself, or in the charge, which the soldier cannot himself rectify, but which in the former case the armourer can and ought to attend to. His officers and non-commissioned officers are responsible that all necessary repairs are made when reported to them, but they cannot be expected to know the defects of every soldier's firelock; this the soldier alone can detect; but not unless he has been previously instructed. He may pass his days in the 3rd class, never hitting the mark and despairing of ever improving in his practice, whilst the fault may really lie in the arm itself or in some other defect of construction, which a better instructed soldier would detect at once. The instructor must therefore draw every man's attention to the following rules:—

30. Very few arms are accurately sighted as to *elevation*; the marks denoting the height to which the sliding-bar should be raised for different distances are seldom exactly in the right place; the sights are all made in the manufactory to one pattern, whereas some muskets will carry a little high and others low; the soldier therefore must pay attention to each shot; if it goes low he must raise his sliding-bar a little, if high the reverse; when he goes home from the practice ground he should apply to his officer to allow him to make a mark on his sight at the proper place, or get the armourer to do it for him, remembering that he has only three or four shots to fire at each distance, and therefore only so many chances of getting his sight adjusted. *Defects of sighting.*

31. The sights are not always in the proper *line*; if the back sight is too much to the right the musket will carry to the right, if the foresight is to the right it will carry to the left; this defect he should rectify either by aiming in the

B 2

contrary direction, or by getting his sight altered by the armourer.

Improper bore. 32. The firelock may not have a proper *bore*. If the soldier finds that his cartridge invariably rams down hard or is very loose in the barrel, he should not fail to report it immediately; this defect, however, is seldom likely to occur.

Injured barrel. 33. If a soldier suspects that his barrel is bent he should report it immediately; if there should be any dent in the barrel it will very likely burst in his hand:—such defects, however, will seldom occur except through carelessness.

Too strong a pull required to move the trigger. 34. If the trigger pulls too hard it will cause the soldier to alter the direction of the arm whilst firing. This is easily rectified, when necessary, by the armourer.

Fore sight. 35. The soldier should pay attention, in cleaning his arms, never to rub the fore sight against any hard substance which would injure it, either forcing it to one side, or blunting it so much that he would be unable to take a proper aim. In leaning his firelock against a wall he should be careful not to make it rest on the point of the fore sight.

Miss-fires. 36. If the musket frequently misses fire, the soldier should ascertain whether his nipple is in good order; this may frequently occur from the communication hole (that is, the hole by which the explosion of the cap communicates with the charge,) not being of sufficient size. It may also be caused by the screw of the nipple being too long when screwed down, thereby shutting up a part of the communication hole, and preventing the powder from getting into the chamber. The soldier should be careful about this, as the instructor will naturally attribute constant miss-fires to *a dirty firelock* and *an inattentive soldier;* ignorance of the construction of the nipple might by this means bring the soldier to defaulters' drill.

Recoil. 37. The explosion of the powder, at the same time that it sends the bullet out of the barrel, also communicates a certain motion to the arm itself; this is called *recoil*. The bullet quits the muzzle (as has already been explained) in the direction of the line of fire; the recoil takes place exactly in an opposite direction. Now, if the soldier will take a musket and imagine the line of fire or axis produced backwards, he will observe that it will pass above the stock; the stock being bent downwards to enable the firer to look along the barrel, the point of resistance (that is the shoulder of the firer) is therefore beneath the line of recoil, and the result is that the explosion has a tendency to throw the muzzle up, and thereby send the bullet high; the lower the point of resistance the more the firelock will fly up, and it is for this reason that in the "Position drill" the soldier is taught to press the *centre of the heel-plate to the shoulder*, and not the toe of the butt."*

38. When the sun is shining from the left it lightens up the left side of the fore sight and the right side of the notch of the back sight. The soldier in taking aim is apt to be guided by these brilliant spots instead of the real centres of the notches, and the result is that the axis will be directed to the right. When, on the other hand, the sun is on the right he will be liable to aim too much to the left. Effect of the sun.

39. It has been frequently found that the least particle of hard rust in the barrel is sufficient to cause the bullet to leave the grooves and cease to turn with them. It should not be necessary to explain to any man what would be the effect if this were to happen; if a soldier cannot prevent this by *keeping his barrel clean*, he is not fit to be trusted with a rifle. Rust is caused by the joint effect of moisture and air; the surest way, therefore, of preventing rust in the barrel is to keep the bore perfectly dry, and invariably to have the muzzle stopper in and the cock down on the nipple, so as to exclude all air. Effects of rust in the barrel.

40. If a soldier is to be out on piquet, or whenever his firelock is likely to be exposed to rain, which should never be the case if it can be avoided, he should stop up the nipple-hole with grease, and let the cock down upon it; or if there is no grease let him drive a peg of wood into it, and put the cap on; neither the grease nor the peg will in any way impede the action of the cap in igniting the charge, but both will disappear in the explosion. How to prevent the powder in the barrel from getting wet.

41. Damp powder will not send a bullet so far as powder which is perfectly dry, and it is of course more likely to miss fire,—for which reason the soldier cannot pay too much attention to preserving his ammunition dry. Effects of damp on the charge.

42. If, in loading, the soldier observes that there is not sufficient powder in the cartridge, he should, in firing, aim a little high, as a small charge will not send so far as the regulation charge. Inexact measurement.

---

* This should be more particularly explained to the soldier during the "Position drill."

Attention to the cleanliness of the pouches inside.

43. The greatest attention must be paid to the cleanliness of the pouches in the inside, especially when the ammunition is in the pouch loose, so that no dirt or dust may adhere to the greased part of the cartridge, which would cause the bullet to stick in the barrel in loading.

Imperfection in the form of the bullet.

44. It has already been explained what effect any unevenness on the surface of the bullet will produce; nothing is more likely to cause this than hard ramming in loading, which ought never to be necessary.

Melted grease.

45. Whenever the grease round the bullet appears to be melted away, or otherwise removed from the cartridge, the sides of the bullet should be wetted in the mouth before putting it into the barrel; the saliva will serve the purpose of grease for the time being.

Pouring in the charge.

46. The necessity of always loading standing when practicable, and of keeping the barrel perfectly upright, should also be inculcated. When the barrel is inclined, as in loading kneeling, a great portion of the powder sticks in the fouling on the sides of the barrel, and causes difficulty in loading.

Preservation of the ammunition in the pouches.

47. Whenever there is loose ammunition in the pouch, that is, when a packet of ammunition has been broken into, the loose cartridges should be folded up in paper that they may not shake about and become damaged. For the same reason the pouch should always be well packed, and no vacant space ever allowed to remain in it.

Firing at a moving object.

48. If an object fired at is moving, whether it is a man walking or a horse at a gallop, it is obvious that it will pass over a certain distance between the moment of discharge and the time that the bullet reaches it. If it is moving across from left to right, or from right to left, the soldier must aim a little to the front of it, but how much must depend, first, on the pace it is going, and secondly, on the distance of the object, and the consequent time the bullet will take to travel. The soldier must be guided entirely by his own judgment in this matter, as no fixed rule can be laid down.

Necessity of constant practice in judging distance by the eye.

49. In conclusion, the instructor should not fail to impress upon his men the great importance of training themselves to judge distance, without which all the firing at a target is so much waste of time. It has already been shown how necessary it is that the back sight of the firelock should be adjusted to the correct distance; but the soldier cannot do this if he is not thoroughly trained to judge distance by

the eye. It is of no use his being a good shot at a fixed mark if he cannot hit the enemy in the field; this is the object of all training.

50. It has been ascertained by experiment, that if the rifled musket, pattern 1853, be fired with the elevation due to 600 yards at an object 570 yards off, the bullet will strike 2·38 feet above the mark; if the musket be fired with the same elevation at the distance of 630 yards, the bullet will strike 2·54 feet below the mark, showing that any error of 30 yards in the appreciation of distance would, at this range, cause the soldier to strike the figure of a man either in the head or feet, according as the error of appreciation was under or over the correct distance (fig. 11). When firing with the 300 yards sight, the bullet will take as much as 70 yards to fall half the height of a man, owing to the trajectory of 300 yards being less incurvated than that of 600 yards. At 800 and 900 yards, the curve being greater than at either of the above-mentioned distances, the same fall would take place in passing over a much shorter distance, consequently the greater the distance the greater the necessity of knowing it accurately. It is for this reason that none but well-trained soldiers should ever be allowed to fire at such distances as 800, 900, and 1,000 yards, and then only at columns of infantry, whose depth would make up, in some degree, for the mal-appreciation of distance. Thus, in firing at a column (fig. 12) whose depth is 100 yards, if the soldier over estimates the distance of the front rank by 100 yards, although such an error would cause him to miss the front rank, he would, if the ground is level, strike the column in its rear. As the soldier, however well trained, cannot always be certain of his distance, it is preferable, when in the field, to give the first shot an elevation rather under than over the correct one; the shot will then strike the ground before reaching the object, and may possibly hit in its bound, or *ricochet*, as it is called. He should be taught to watch the effect of his shot, which may generally be ascertained by observing the dust thrown up when the bullet strikes the ground; he can then adjust his sliding-bar by raising it higher or lower, according as his first shot strikes before or beyond the object.

51. The foregoing course has been drawn up (as stated in the heading) for the guidance of officers in the instruction of their men; but as it is necessary in lecturing to the soldier to avoid as much as possible all words that may not be understood by the least educated man present, it has been thought advisable to append a list of some of the words therein contained, with the simplest explanation of the sense which they are intended to convey.

Adhere.—To stick to.

Adjustment.—The act of regulating; thus, when a sight is set to a correct height it is said to be adjusted.

Angle.— A point where two lines meet. Angles are of different sizes, but the size does not depend on the *length* of the lines which form it, but on their *direction* with regard to each other; thus, if a common carpenter's rule with a single hinge be opened, the two sides are said to form an angle with one another, and the more it is opened the greater the angle will be, until at last it ends in a straight line.

Aperture.—An opening,—a hole running a short way into the back of the bullet.

Applicable.—Suitable, relating to.

Ascertain.—To make certain, establish, find out.

Assigned.—To mark out; to assign a distance of 200 yards to any place is to suppose that it is 200 yards to that place.

Atmosphere. — The air that encompasses the earth on all sides.

Attribute.—To lay to the charge of.

Circumference.—The outside of a circle or ball.

Construction.—Form, shape.

Contact.—Touch; any two substances touching one another are said to be in contact.

Convex.—Rising in a circular form, as the outside of a circle or curve.

Cylindrical.—Long and round like a roller; that part of the bullet which is of this shape is said to be cylindrical.

Defective.—Imperfect, of bad shape.

Denoting.—Meaning.

Diameter.—The line drawn across a circle or ball from one side to the other, passing through the centre, and dividing it into two equal parts.

Dilating.—Widening, extending.

Direction.—With regard to firing, is in contra-distinction to *elevation*, and is taken to express the course which a bullet takes to the right or left; thus, if a bullet strikes the ground short, but in the same line as the bull's eye, the *direction* is said to be good, but the elevation insufficient; if, on the other hand, it hits the butt on the same level as the bull's eye, but to the right of it, the elevation is said to be good, but the direction defective.

Diverted.—Turned aside, or drawn away from.

Elevation.—The angle formed between the line of sight and line of fire; this is sometimes called angle of sight; properly the elevation should be taken to imply the angle formed between the line of fire and the horizon called angle of fire, but it is often received in the first-mentioned sense, the line of sight being usually horizontal when at target practice.

Elongated.—Lengthened, drawn out.

Expanded.—Spread, enlarged, dilated.

Experiment.—Trial, proof.

Explosion.—The act of driving out with force and violence.

Excess.—That which is over and above; excess of windage is that which is too much.

Groove.—Furrow, indentation.

Horizontal.—Level; the surface of smooth water is horizontal.

Inclination.—Leaning, bending; a sight is said to be inclined if it is not perfectly upright when held at the "Present."

Initial.—First, beginning, original.

Lateral.—Growing out on the sides.

Moulded.—Modelled, shaped, formed into a shape.

Musketry.—This word is now taken to embrace all that relates to the construction or employment of the musket, and instruction in its use. The school of musketry implies school at which everything relating to the musket is taught.

Obviated.—Prevented.

Particles.—Small portions of matter.

Perceptible.—Distinguishable, observable.

Previously.—Beforehand.

Produce.—When a line is lengthened in the same direction it is said to be produced.

Rectified.—Set to rights.
The Reverse.—The contrary, exactly the opposite.
Range.—The distance to which a bullet will fly with any given degree of elevation.
Resistance.—Opposition, acting against.
Rotatory.—Whirling, as a wheel turning round.
Reaction.—The force imparted to any body by its contact with another.
Ricochet.—The bounding of a musket bullet or cannon shot along the earth.
Section.—A representation of anything cut through, so as to show the shape of it inside. A distinct part of a book or writing.
Spherical.—Round.
Tending.—Calculated to move towards.
Traversing.—Moveable, going across from one side to another.
Velocity.—Swiftness of motion.
Vertical.—Upright; if a plummet is hung from a string its direction will be vertical.

PART III.

Preliminary Instruction in Firing.

TARGET DRILL.

1. For this exercise the traversing rest must be used to support the firelock; or three stakes tied near the top, and supporting a bag of sand about 4½ feet from the ground, will answer the same purpose. Aiming with a rest.

2. A squad should never exceed ten men at a time at each *rest;* it will be formed in a single rank, each man having his own firelock. The instructor will first explain the principles of aligning the sights of a musket on an object, confining the attention of the squad to the following simple rules:—

   1st. That the sights should not incline to the right or left.

   2nd. That the line of sight should be taken along the centre of the notch of the back sight and the top of the fore sight, which should cover the middle of the mark aimed at.

   3rd. That the eye should be fixed steadfastly on the mark aimed at, and not on the barrel or fore sight, which latter will be easily brought into the alignment if the eye is fixed as directed. Particular attention must be paid to this rule, for beginners are apt to fix the eye on the fore sight instead of the mark, in which case the latter can never be distinctly seen, and the difficulty of aiming is greatly increased.

   4th. That in aiming, the left eye should be closed; if any of the squad are not able to do so at first, they will soon succeed by tying a handkerchief over the left eye.

3. The instructor will then explain the difference between *fine* and *full* sight in aiming; the former being when the line of sight is taken along the bottom of the notch of the back sight, the fine point of the fore sight being only seen in the alignment as A : the latter is when the point of the fore sight is taken in alignment with the shoulder of the notch of the back sight, as B.

4. As these two methods of aiming cause a slight difference in the angle of elevation, it is necessary the soldier should understand that the ordinary rules for aiming are intended to apply to *half sight*, which means that the alignment is taken with the summit of the fore sight at half distance between the shoulder and bottom of back sight, as C.

5. As some firelocks will carry higher, and others lower than the average, allowance can be made for this defect by aiming with full sight when the musket is found to carry low, and by aiming with fine sight when it carries high; when, however, no such defect is observed in the practice with the firelock, the men are invariably to be taught to aim at half sight.

6. Having explained the foregoing rules, the instructor will cause each soldier to take aim at an object of the same size as the bull's eye used in practice, at every distance of fifty yards from 100 to 900 yards, viz.:—

From 100 to 300 yds.—Bull's-eye, 8 in. in diameter.
From 350 to 600 yds.   „   2 ft.   „
From 650 to 900 yds.   „   4 ft.   „

7. After each man aims he will step aside, in order that the instructor may examine and see if the aim is correctly taken; should he observe any error, he will cause the next man to advance and point out the defect; the error, however, is always to be corrected by the man who has aimed.

8. To vary the practice, the squad should occasionally be exercised at intermediate distances (as 425 yards for example), as also be made to aim at a soldier placed in front of the target, or at a group of several men together.

POSITION DRILL.

9. The *position drill* differs from the *platoon exercise;* the latter comprehending the positions of loading and firing in the ranks, in which the soldier is instructed by the Adjutant, whereas in the "position drill" the attention of the instructor of musketry is to be confined exclusively to the essentials of good independent firing.

10. For this drill the squad (which should never exceed 10 men) will parade in marching order, and be formed in single rank at one pace apart, and placed at any convenient distance from the target or mark; the instructor will then order the squad to fix bayonets and proceed with the

" position drill," first in slow time *standing*, according to the instructions hereafter detailed ; and as too much pains cannot be taken to ensure that each man takes a deliberate aim at some specified object *whenever he brings his firelock to the " present*," if no natural object presents itself for the man to aim at, several small bull's-eyes must be marked on the barrack wall,—

1st. Load.—According to regulation.

2d. Ready.—Adjust the sight and proceed according to regulation.

3d. Present.—Bring the firelock at once to the shoulder, pressing the centre part of the heel plate firmly into the hollow of it with the left hand, which must grasp the piece at the " *swell*," the right hand holding it at the " *small*," the right elbow raised (but, when firing in platoon, not so much as to impede the aim of the rear rank man), the muzzle inclining to the bottom of the object and the forefinger of the right hand extended along the side of the trigger guard, at the same time shut the left eye.

" Two."—Raise the muzzle steadily until the foresight is aligned through the back sight with the object the right eye is fixed upon, placing the second joint of the fore finger on the trigger at the same time, and restrain the breathing.

" Three."—Pull the trigger with the second joint of the fore finger, by a steady pressure, without the slightest jerk or motion of hand or elbow, keeping the eye still fixed on the object, as in the preceding motion.

" Four."—Bring the firelock down to the capping position and shut down the flap, at the same time bring the right foot to the position in which it was placed before coming to the ready ; count a pause of slow time, and come to the position of prepare to load.

4th. Load.—According to regulation.

11. Having thus put the whole squad through the drill in slow time, and corrected the position of each man, the instructor will order it to continue the motions of loading and firing *independently*, each man aiming at a specified mark. The most minute attention is to be given to each man's position when at the " present," and more especially that the firelock is pressed firmly to the shoulder with the left arm, and that the trigger is pulled without the slightest jerk, and with the motion of the fore finger only, the eye being fixed upon the mark during and after snapping the lock. In this drill the instructor should frequently place himself in front of the squad at five or six paces distant, and cause each man to aim at his right eye, in order to ascertain that he obtains the alignment quickly and correctly, and that his aim is not disarranged by pulling the trigger; this is of the utmost importance.

12. When the men have been sufficiently exercised in the position of firing standing, they will be put through the drill in the kneeling position with unfixed bayonets, going through it at first at slow time, according to regulation, observing the several points to which the attention is called in the foregoing remarks.

Judging Distance Drill.

Instruction for Recruits and others in the Company.

13. In order to apply the rules of firing laid down for the musket, it is necessary to know the distance which separates a man from the object he is firing at.

14. In firing for instruction, the target is generally placed at known and measured distances, but before the enemy the distance is unknown; it is necessary therefore to judge the distance quickly and exactly, in order to regulate the elevation of the piece accordingly.

15. In order to teach the soldier to estimate distances by the eye, he will be instructed according to the following rules in the first instance, before he passes on to the method contained in the "*Judging Distance Practice.*"

16. The instructor will cause a line of 300 yards to be measured accurately; this line will be divided into equal parts of 50 yards each, by perpendicular lines of the length shown in the diagram (pl. 4).

17. At the extremity of each of these perpendicular lines, the instructor will place a soldier standing at ease, and facing the squad he is about to instruct.

18. It will be observed, that each of these soldiers is placed at a greater distance from the line of 300 yards, in proportion as he is distant from the point where the squad will commence their instruction, in order that each soldier may serve in turn as a point of distance for the squad to make observations on.

19. The instructor will point out successively to the men the different parts of the figure, arms, accoutrements, and dress, which they can still perceive distinctly on the soldier placed at 50 yards distant, and also those parts that they can no longer perceive clearly at this distance; he will question the men one after the other in the observations they make on what they see, but he must not expect that the answers should be the same from every man, since the eye-sight is not the same in all. Every soldier will try to impress upon his mind the appearance of the man placed at 50 yards.

20. The instructor then, by moving the squad to the right, will place it in front of the soldier at 100 yards distant, and cause each man to make observations of the same kind as on the man at 50 yards, and desire him to make comparisons between the two men placed at these two distances.

21. The instructor will then pass on to the other distances, proceeding in the same manner as for the first two.

22. He will endeavour above all to point out to each soldier, according to the observations he may make, the differences that exist between the men placed at the six different distances comprised in the sub-divisions of 300 yards, pointing out at each distance what parts of the figure, dress, and equipment are clearly perceivable, those that are seen confusedly, and those that are no longer visible.

23. The instructor will cause the men to take notice of the position of the sun, and state of the atmosphere and back ground at the time they are making their observations, in order that they may be accustomed to the alterations made in this respect in the appearance of the several objects.

24. The men who are placed as points will then be relieved, for which purpose the squad will consist of at least double the number of men employed as points.

25. When all the men of the squad shall have made a sufficient number of observations on the different points designated, and when these observations are well engraved on their memory, the instructor will proceed in the follow-

ing manner to the estimation of distances comprised within the limits of 300 yards.

26. After having marched the squad on to different ground from that on which the appreciation of distances has taken place before, the instructor will form them in single rank, and will send a man to the front, marching him by means of the bugle (if there is one) diagonally to the right and left, and occasionally at the double, in order that the rest of the squad may not count his paces; then at any convenient distance within 300 yards, he will command "Halt," when the man will face the squad, and "stand at ease." He will then order the men to observe the soldier who is facing them, and to estimate the distance, recollecting the observations they have previously made on the men placed at measured distances.

27. The instructor will then call each man separately to the front and question him, noting down his answer,— which must be given in a low tone of voice, in order that those following him may not be influenced by his opinion.

28. Every man will adjust the sight of his firelock to the distance he judged.

29. When all the men have given their answers, the squad will proceed to measure the correct distance by advancing towards the man judged from, the instructor placing himself in the centre, the men counting the number of paces, the instructor only counting them aloud.

30. The men should be taught to measure the distance in the following manner, at every 120 paces they will double up one finger of the right hand to mark 100 yards, commencing again 1, 2, 3, and so on. When at the end of any division of 100 yards the remaining distance appears to be within 100 yards they will commence counting by *tens of yards* by doubling up a finger at every twelve paces. The correct distance will however be ascertained by actual measurement with a cord or chain, for which purpose two or three men will follow immediately in rear of the squad.

31. The instructor in repeating this exercise will take care that as much as possible it is conducted in different directions, and under different states of the atmosphere, in order that the soldier may become habituated to the diversity of circumstances in which he may have to act.

32. The men, after they have been drilled up to 300 yards, will continue the exercise up to 600 yards, first at fixed points at every 50 yards from 350 to 600 yards, and

afterwards at unknown distances. In exercising the men at great distances, it will be desirable to separate the squad into two equal parts, facing each other. After every man has judged the distance which separates them, they will advance towards one another, each party measuring half the distance; by this means much time and walking are saved.

33. The number of drills to be devoted to this exercise will be arranged as follows:

Four drills, at fixed points to 300 yards.
Three ,, at unknown distances up to 300 yards, each drill to consist of four answers.
Two ,, at fixed points from 300 to 600 yards.
Three ,, at unknown distances, from 300 to 600 yards, each drill to consist of four answers.

MANUFACTURE OF CARTRIDGES.

34. Articles for the instruction of soldiers in the manufacture of cartridges will be supplied to each barrack by the war department, according to the list at page 58.

35. To construct the cartridge, cut the paper according to the patterns (plate 5); place the rectangle on the little trapezium, the sides A, B, C, of the rectangle coinciding with the sides A, B, C, of the trapezium; lay the mandrel on the *rectangle* parallel to the side B, C, the base of the "*mandrel*," even with the side C, D, of the *rectangle*; roll the whole tightly on the "*mandrel*," place it vertically, and fold the remainder of the *trapezium paper* into the hollow in the base of the "*mandrel*," commencing with the acute angle of the *trapezium*; make use of the point of the "*former*" to close the folds; examine the bottom of the inner case thus formed to see that there remains no hole for the escape of the powder when charged. Introduce the point of the bullet into the aperture at the base of the "*mandrel;*" take the *trapezium envelope,* and place the mandrel and bullet parallel to the side F, G, with the bottom of the bullet at half an inch from the base F, H, of the envelope; press up the point of the bullet into the cavity, and roll the envelope tightly on the bullet and on the *mandrel;* fold the remainder of the envelope on the base of the bullet, commencing with the acute angle; place the base of the cartridge on the table, and withdraw the *mandrel* by squeezing the case of the cartridge with the left hand and raising up the *mandrel* with the right.

c

36. To charge the cartridge, introduce the point of the *funnel* into the bottom of the case of the cartridge; pour in 2½ drams of fine sand from the *powder flask ;* then withdraw the *funnel,* taking care that none of the sand escapes between the case and the envelope ; squeeze the top of the cartridge and twist it round.

37. When completed, the base of the cartridge must be dipped up to the shoulder of the bullet in a pot of grease, consisting of six parts of tallow to one of bees' wax.

Target Practice.

38. The targets for this practice will be six feet in height by two in breadth ; they will be constructed of cast iron, three quarters of an inch thick, and squares of six inches cut on the face to facilitate the marking off of the hits in the diagrams provided for the purpose; in the centre is a bull's eye, eight inches in diameter ; and from the same centre, with a radius of one foot, a black circle is described, dividing the target into two parts, centre and outer (plate 6). Circular rings will also be cut on the face of the target, to serve as guides in painting it.

39. The white part of the target should be white-washed, and a pot of white-wash, together with a pot of colouring for the bull's eye and black circle, should be kept in rear of the target.* The shot will be found to make a very distinct mark on striking the face of the target, so that the correct position of each shot is easily distinguishable at a short distance.

40. In all cases when the nature of the ground admits of it, a trench should be dug for the markers, of the dimensions given in plate 7, about 15 yards to the front, and to one side of the targets, and in such a position that the markers may easily see the face of the target from it; the earth excavated should be thrown up on the side of the firing ; there should also be two epaulments, so as to screen the men, not only from the shots themselves, but from any stones that may be thrown up by them.

41. A fatigue party will fix the targets before the practice commences, and lay down the cord or chain for the judging distance practice, and whilst the men are firing, they will assist in marking, or in any other duty that may be required.

---

* It has been found that the colouring which answers best is made by mixing whiting or lamp black with water and size.

42. Two men of the fatigue party will, when necessary, be placed as sentries to the right and left of the butt, and clear of the range, to prevent any persons from passing within the line of fire.

43. Each man will expend, as his annual allowance of ammunition, 90 rounds in the following manner, viz., 60 in firing individually; 10 in file firing and volleys; and 20 in firing in extended order.

44. The targets will be arranged as follows for the different distances:—

Up to 200 yards the practice will be at a single target.
At 250 and 300 yards - ditto - at 2 targets.
,, 350 ,, 400 ,, - ditto - ,, 3 ,,
,, 450 ,, 500 ,, - ditto - ,, 4 ,,
,, 550 ,, 600 ,, - ditto - ,, 5 ,,
,, 650 ,, 700 ,, - ditto - ,, 6 ,,
,, 750 ,, 800 ,, - ditto - ,, 7 ,,
,, 850 ,, 900 ,, - ditto - ,, 8 ,,

45. The troops will fire at every distance of 50 yards from 100 to 900 yards. These distances are divided into three parts, viz., up to 300 yards included, will be for the practice of the 3rd class, as far as 600 yards included for the 2nd class, and the 1st class only will continue the practice to 900 yards.

46. Target practice will invariably take place in marching order.

Individual Firing.

47. The company or class will be marched to the ground in open column of sections right in front, and halted so that the right of the leading section may rest on the point selected to commence the firing, the column facing the targets; the remaining sections will then open out to double distance from the front; pile arms and take off their knapsacks, placing them in a line in rear of their arms.

48. The non-commissioned officers and men who have been previously told off as markers, will then be sent to the marker's butt, to mark, and give the established signals, which will be denoted by flags of different colours, to be raised above the butt, as the shot strikes.

49. Ricochets, or shots which strike the ground first before they strike the target, are to receive no signal, and are to be counted as misses in individual firing.

50. The signals for the different distances, and the value attached to each shot, will be as follows:—

	Shots.	Flags.	Value.
In the practice of the 3rd class.	Outer	White	1
	Centre	Dark blue	2
	Bull's eye	Red and white	3
	Miss		0
Practice of the 1st and 2nd classes.	Outer	White	1
	Centre	Dark blue	2
	Miss		0

51. The danger or cease-firing signal will in all cases be a red flag. This will be hoisted whenever it is necessary to "cease firing," in order to run out to wash the target or for any other purpose; it will invariably be answered from the firing point by sounding the "cease fire," and will always be kept up as long as the markers are out of the butt. Whenever the "cease fire" is sounded from the firing point, it will be answered by raising the danger flag from the marker's butt, and in like manner the "commence firing" will be answered by lowering it.

52. Whenever a shot strikes the target to the right, the flag denoting the value of the shot will be inclined to the right, and *vice versâ*; when the shot strikes high, the flag will be raised as high as possible; and when low it will only be raised high enough to be easily distinguishable above the butt.

53. It is to be understood, that in all practices, whenever a shot strikes the target, so that the circumference of the mark made by it, cuts within the circumference of the bull's eye or centre, such shot is to be counted as hitting the bull's eye or centre; and the circumference of these divisions is in all cases to be taken to the *outer* edge of the mark cut on the face of the target.

54. The non-commissioned officer-instructor of the class or company will keep a register of the form marked B (plate 9); on this he will note, under the number of the shot fired, the value or number of points obtained by it, whether 1, 2, 3, or 0. At the conclusion of the practice, he will add up the total number of points obtained by each man during the practice; the addition of the column of total points will give the total of the squad or section, and this divided by the number of men will give the average, should it be required.

55. All entries are to be invariably made *in ink* on the ground; and should any erasures be necessary, a fine line will be drawn through the figure thus— the correction made, and the officer's initials immediately attached to it. | 2 1 *A.L.F.* |

56. The men's names are to be written in the register before the party comes out, in the same order as in the Company's Practice Return, and according to which they will stand in the ranks for firing; in general, one register will suffice for each section.

57. The marker in the butt is invariably to be a non-commissioned officer of a different company from that engaged in firing, he will be responsible for the correct signals being given to the several shots which strike the target, and will, if convenient, keep a diagram of the practice as it proceeds. Should it not be considered necessary to keep a diagram (which will be entirely left to the judgment of the instructor) the marker will keep a memorandum of each shot as it strikes, under the head of bull's eye, centre, and outer. This will prevent delay and ensure each man's shot receiving the correct signal.

58. When the party or section has loaded by word of command, and everything is ready to commence, the bugler, who is placed on the right of the firing point, will sound the "commence firing," and after the danger signal has been lowered, the officer will order the right-hand man of the front rank to go on. After he has fired he will immediately fall three paces to the rear, having previously come to the *shoulder* from the "capping position;" the next man of the front rank will then move to the firing point and fire, after which he will also fall three paces to the rear of the point he previously occupied. In like manner every man of the front rank will fire in succession, after which the rear rank will commence on the right, and after firing they will form in rear of the front rank, so that by the time the whole section has fired one round, it will have re-formed three paces in rear of its original position.

59. The non-commissioned officer of the section will then advance his section three paces, and load, after which the firing will proceed as before.

60. The non-commissioned officers of each section will fire at the head of their sections according to seniority, and the company-instructor should fire at the head of his company or class.

61. The instructor will take care not to correct a man at the moment he is firing, which would produce no other effect than to distract his attention from the object he is aiming at; but will observe attentively the position of each soldier, and correct him after he has fired.

62. Whenever the hits on the target become too numerous to distinguish quite easily the fresh ones as they strike, the target should be whitewashed afresh; before which, however, the company-instructor will carefully compare his register with the target; the subsequent hits will then be marked on the diagram (should one be kept) with a +, or some other mark to denote the shots which have struck the target after it was whitewashed.

63. In the practice *as a company* in the third class, the fourth section may be practised in judging distance, whilst the first section is firing, and each section after it has fired may proceed to the judging distance practice. When, however, it is found not convenient to carry on the target and judging distance practices at the same time, the men who are not firing should always be occupied in the aiming and position drill.

64. All those men who are not occupied in the above-mentioned exercises, and who desire to watch the practice of their companies, are to stand on the right of the firing point; they are always to be kept clear of the section that is firing, and on no account is any noise or talking to be allowed between them.

65. At the conclusion of each practice, the bugler will sound the assembly, when the company-instructor will proceed to compare the register with the target, as also the diagram (if one has been kept); both register and diagram will then be completed and signed by the *company-instructor*, and *marker*, and countersigned by the *officer-instructor*, after which the " *duplicate total points* " initialed by the officer-instructor, to verify its agreement with the column "*total points*," will be torn off and immediately handed over to the non-commissioned-officer instructor of the battalion, who on the practice ground is especially responsible that this order is rigidly attended to in all cases.

66. The company-instructor will, immediately on his return to barracks after every practice, transcribe the column "total points" of the registers to the company target practice return F.

67. When there are *casuals*, who have to make up their practice on a subsequent day, the register and diagram will only receive the initials of the officer-instructor and serjeants, and the "*duplicate total points*" will not be filled in; these documents will be then given over to the non-commissioned-officer instructor of the battalion, who will take charge of them until required.

68. It is desirable that every company, or section of a company, or class, should be made as complete as possible whenever it goes to target practice, to prevent the inconvenience and delay arising from *casuals*. Should, however, any men be absent from an unavoidable cause, when a company, section of company, or class is firing, the whole of such men, termed "casuals," must parade at *the same time*, and, if possible, when the company or class next goes out to practice, in order to complete the distances missed; but should it be impossible to do so, they must parade on a subsequent day, in which case a second column of casuals will be ruled in the register, to record the total of their practice. Any men who may have missed firing at two distances, and are not present with their company, section, or class the next time it proceeds to practice, are not to be further exercised in the "period" of practice in which the company or class is engaged; the register and diagram of the section or class to which the *casuals* belong for the distances missed will be taken out, and their total practice recorded under the column of casuals; the hits will be marked in the diagram with some sign, to distinguish them from the shots previously obtained by their section or class. When completed, the register and the diagram will be signed in full, and the columns of *casuals* will be totaled up; the total of these columns, added to the column "total points," will make the column "duplicate total points," which, being found correct, will be initialed and torn off, and disposed of as before directed.

69. The practice of individual firing is divided into three "periods," in each of which the soldier will fire twenty rounds.

First Period.

70. The battalion will commence the "*first period*" yearly with the practice of the 3rd class, which will be carried on by companies under the command of their

Practice of the company in the third class.

captains, superintended by the officer-instructor. All shots which hit the bull's eye will be marked with No. 3 in the register; those that strike the centre with No. 2; those that strike the outer with No. 1; and the misses with 0. Each man will fire four rounds at 100, 150, 200, 250, and 300 yards. At 100 and 150 yards the practice will be conducted standing and with fixed bayonets, and at the other distances with unfixed bayonets, the men being allowed to stand or kneel at pleasure.

71. When the whole of the company has executed the practice up to 300 yards in the 3rd class, the total points obtained individually at each distance in the "first period" will be added together, to show the practice of each man in the 3rd class in the company practice return marked F. From this column the company will be divided into two classes; all men who have obtained in this practice 13 points will pass into the 2nd class, the remainder will recommence the practice of the 3rd class at 100 yards. This "period" of the return will be signed by the captain of the company, as a proof of its correctness, and by the officer-instructor after he has carefully examined and compared it with the "duplicate total points" in his possession.

72. The names of the men who have passed into the 2nd class will be read to the companies on parade by their captain.

SECOND PERIOD.

Practice of the second and third classes.

73. After the men of the company have been divided into classes, the practice will no longer continue as a company, but be carried on by classes under the superintendence of the officer-instructor of the battalion. Each class, if the number will admit of it, will be divided into sections, and their names will be placed in the registers of each company in the order in which they stand in the practice return.

74. The company instructor will attend when possible with every class.

75. Whenever there is a choice of time for practice, the senior class will always have the advantage.

76. The third class will repeat the practice from 100 to 300 yards, firing as before directed.

77. The second class will fire three rounds per man at the distances of 350, 400, 450, 500, and 550 yards, and five rounds at 600 yards.

78. In the practice of the second class, all shots which strike the centre will be marked No. 2 in the register; those which strike the outer with No. 1, and the misses 0. The bull's eye in this practice will only count as centre, the whole of which is to be painted black.

79. The practice of the second class will be carried on throughout with unfixed bayonets, the men standing or kneeling at pleasure.

80. At the conclusion of the practice of the "second period," the company instructor will total the points obtained at each distance by the men of the second and third classes recorded in the practice returns, which the captain of the company will then sign and send to the officer-instructor of the battalion, who, having found that it is correct, will also attach his signature and return it to the company.

81. A second classification will now be made, when all those men who, in the practice of the second class, have obtained ten points will pass into the first class; the remainder will repeat the practice of the second class.

82. The qualification for passing from the third to the second class will be the same as in the first period.

THIRD PERIOD.

83. The three classes will be told off into sections as before; the second class will now be composed partly of men who repeat the practice of the second class, and partly of men who have passed out of the third class in the second period. *Practice of the first, second, and third classes.*

84. The practice will be conducted on the same principles, and the hits will have the same value, as in the second period, except that the centre in the practice of the first class will have a diameter of four feet instead of two feet, all of which to be painted black.

85. The first class will fire three rounds per man at the distances of 650, 700, 750, 800, and 850 yards, and five rounds at 900 yards.

86. At the conclusion of this period the columns of the "third period" in the company's practice return will be totaled, and from which a final classification for the year will be made. From this the company instructor will make out a list, in which each man will be placed according to his performance, with the number of points obtained in the third period attached to his name, viz.,

those who obtain most points in their class first, and so on; which list will be posted up in the barracks.

87. The men of the first class will be exempt throughout the following year from target drill.

88. That man who obtains the greatest number of points in the practice of the first class will receive the prize as best shot of his battalion.

89. Should two or more men obtain the same number of points in the practice of the first class, the prize will be awarded to that man who has obtained the greatest number of points throughout the whole practice of individual firing.

File and Volley Firing.

90. This practice will be carried on as a company under the command of the captain, the men of all classes being united, and firing by sections two deep, in marching order and with fixed bayonets, at the distance of 300 yards, each man expending five rounds in "file firing," and five in "volleys," of which two will be delivered as a company kneeling.

91. The mark will consist of eight targets placed close together, each having a separate bull's-eye and centre. The shots which strike the targets will be valued as in the third class.

92. At the conclusion of the practice of each section, the company instructor and a serjeant of another company, together with the non-commissioned-officer instructor of the battalion, will go up to the targets and mark the hits on a diagram, which being completed will be signed as in the preceding practices by the two serjeants and by the officer-instructor of the battalion. The diagram (which in this practice is not to be dispensed with), will then be immediately given over to the non-commissioned-officer instructor of the battalion.

93. The company instructor will keep a memorandum of the total points obtained in this practice and insert them in the proper place in the practice return; this column will bear the signature of the captain and officer-instructor of the battalion.

Skirmishing.

94. This practice will also be carried on as a company, by sections, under the command of its captain. It will be

divided into two practices, in each of which ten rounds per man will be expended in extended order at six paces.

95. Eight targets, each having its bull's-eye and centre, will be placed with intervals of six paces between each. Every file to have its own target, and the hits will be counted as in the practice of the first and second classes.

96. In the first practice, five rounds will be fired kneeling at 300 yards, and five in advancing to 100 yards; on the section arriving within 100 yards of the targets, should the ammunition not be expended, the "cease fire" and "retire" will be commanded, and when it has moved to the rear, 250 yards from the targets, it will again be ordered to advance and fire until the remaining rounds are expended.

97. In the second practice, the ten rounds will be fired advancing and retiring between 200 and 400 yards, each man judging his distance and arranging his sight accordingly.

98. The instructor in this practice will take care that the men of the third class, who have not fired beyond 300 yards, arrange their sights to the proper elevation.

99. In firing advancing the men will be allowed to kneel to fire, rising to load.

100. The sentries placed on each flank of the butt to keep the ground will prevent any person from approaching within fifty yards of either flank of the line of targets.

101. At the conclusion of each practice a diagram will be marked off, recording the shooting of each section, and when completed, signed as directed in the preceding practices, and immediately given to the non-commissioned-officer instructor of the battalion.

102. The company instructor will make a memorandum of the points obtained by each section in each practice, the totals of which will be inserted in the proper place of the practice return, and bear the signatures of the captain and officer-instructor.

103. The average points obtained in these practices, added to the average obtained in the "*practice of the company in the third class,*" and that of "*file-firing,*" and "*volleys,*" will denote the merit of the company; and that company which has the highest figure will be the best shooting company in the battalion. No man is to fire either in the "file and volley" or "skirmishing" practice, who has not fired up to 300 yards.

REMARKS.

104. When the number of rounds available for the yearly course of target practice is not sufficient to carry on the practice as detailed in the above, the following alteration will be observed in the order of firing :—

- 1st. When only 70 rounds per man are available, the practice will be conducted as detailed in these instructions, except that the practice of the third period will be omitted. Should there be any number of rounds over, they will be used in the instruction of the men of the third class, by firing from a rest at distances from 300 to 600 yards.
- 2nd. When 60 rounds, the practice of the third period will be omitted, and 10 rounds only will be expended in the second period; viz., in the third class, 5 at 200 and 5 at 300 yards, and in the second class, 5 at 400 and 5 at 600 yards.
- 3rd. Whenever the practice ground does not afford a longer range than 600 yards, and 90 rounds per man are available, the men of the first class will expend the 20 rounds allotted to the practice of the first class in skirmishing, advancing, and retiring between 600 and 400 yards. This practice will be registered, but not included in any return.

105. When the range only extends to 400 and 500 yards, all those men who pass into the second class during the first period will expend the amount of ammunition allotted to the second period at distances between 300 yards and the extreme limits of the range, taking care that the ammunition is equally divided between these distances. The practice of the third period will be omitted, and the battalion will not expend more than 70 rounds per man in the annual course of practice.

106. When the range extends to 300 yards only, all those men who pass into the second class during the first period will repeat the practice of the third class during the second period. It must be understood, however, that when a man has once passed into the second class, he cannot be reduced during the annual course of practice, even although he may fail to obtain the number of points necessary to pass to a higher class, in going over the same distances a second time. At the close of the second period, therefore, no further classification can be made in

the case of those who have already passed into the second class in the practice of the first period, but a re-organization of the classes will take place, when every man will be placed individually according to his performances in the second period. The skirmishing in this case can only be carried on between 300 and 200 yards. The battalion will only expend 70 rounds during its annual course of practice.

107. So long as 50 rounds per man are available, and the range extends to 300 yards, the following practices will in all cases be gone through without the slightest alteration either of the amount of ammunition or distance, viz., "*the practice of the company in the third class,*" "*file-firing and volleys,*" and "*skirmishing practices.*"

108. Whenever a single company is detached from its head quarters, and means of practice are provided, the company will exercise in strict conformity to the instructions herein contained, and at the conclusion of the practice, a company return of the established form will be sent in to head quarters. An assistant non-commissioned-officer instructor should be temporarily appointed in addition to the company instructor.

109. Whenever two or more companies are detached, the assistant officer-instructor should join the detachment during the period of target practice; provided always, that both regiment and detachment are in the United Kingdom, or in any of Her Majesty's dominions where this may be practicable.

JUDGING DISTANCE PRACTICE.

110. The following course of judging distance practice will be gone through annually by every soldier of the battalion, and, when possible, will be carried on by the sections not occupied in firing, when at target practice.

111. A cord or chain of the length required for the practice (divided into parts of five yards each, with the distances of each division from the end so marked as to be distinguished only on close inspection,) to be stretched in any direction that may be found convenient for the practice, care being taken to vary the ground as much as possible for the several practices.

112. One or more men, when judging at 300 yards only, but beyond that distance a section of not less than eight or

ten file, will be stationed at the end, or at any other part of the chain that may be directed, to serve as objects to estimate from.

113. The answers of each man to be recorded in a register of the form marked C, which will invariably be kept by a non-commissioned officer of a different company to that under exercise.

114. The strictest silence is to be observed throughout the practice,—the men are to be prevented from consulting together in judging their distance,—and, in giving their answers, must speak in a low tone of voice so that they may not influence in any way the judgment of each other.

115. The commander will fix on a point at any uncertain distance to commence the practice, to which he will march the section or party, halting at about ten paces either to the right or left, and facing the objects; he will then arrange the non-commissioned officers who are to keep the register three paces to the front of their several sections, to prevent if possible the answers, when given, being heard by those in rear; these non-commissioned officers will then call in succession upon each man of their sections, who will be required to judge the distance in yards, and give his answer, which will be immediately noted down in the register.

116. As the commander will always select a division of five yards at which to halt the section or party, the men must be cautioned to complete a division of five yards in giving their answers.

117. When all the answers of each section or party have been noted down they will be read over to the men, and any error which may be discovered will at once be corrected; after which the commander will refer to the chain or cord, and state aloud to the men the correct distance, which will at once be noted at the top of the column, the number of points obtained by each individual being at the same time registered and made known.

118. In each practice the men will be exercised at four different stations. When the section or party has been exercised at one station it will be moved to another in a manner that will preclude the possibility of any clue to the actual distance being obtained.

119. At the conclusion of each practice the number of points obtained by each man will be read over to the men; and the register when completed by filling up the

column "total points" and "duplicate total points," (which is always to be done on the practice ground,) will be signed by the non-commissioned officer who has kept it, and by a non-commissioned officer of the company exercising, and countersigned by the officer-instructor, who will also place his initials to the "duplicate total points," which are then to be torn off and given over to the non-commissioned-officer instructor of the battalion; the company instructor retaining the register, the total points of which he will invariably transcribe into the company's judging distance practice return immediately on his return to barracks.

120. When there are casuals, the column duplicate total points will not be filled up in the register, which will only receive the initials of the parties before-named, and be handed over to the non-commissioned-officer instructor of the battalion.

121. When the casuals are to make up their judging distance, the company instructor will go to the non-commissioned-officer instructor of the battalion for the register of the section to which they belong, and after the same has been completed, the signatures will be attached in full, the register will then be kept by the company instructor, the non-commissioned-officer instructor of the battalion receiving the duplicate total points.

122. Any casuals who are in arrear more than two practices will not be further exercised in the "period" in which their company or class is engaged.

123. The practice of judging distance, like the target practice, will be divided into three periods, each period consisting of four practices. The third class will practise as far as 300 yards, the second to 600 yards, and the first as far as 900 yards.

124. The value of the men's answers by points in the several classes in judging distance will be registered as follows:—

3rd class, or when judging distance between 100 and 300 yards. } Within 5 yards, 3 points.
,, 10 ,, 2 ,,
,, 15 ,, 1 ,,

2nd class, or when judging distance between 300 and 600 yards. } Within 20 yards, 2 points.
,, 30 ,, 1 ,,

1st class, or when judging distance between 600 and 900 yards. } Within 30 yards, 2 points.
,, 40 ,, 1 ,,

125. It is to be observed that should the first or second class be brought to judge within the distance of an inferior class, which, in order to test the proficiency of the men, ought frequently to be done, the points should then only be counted agreeably to the conditions laid down for these classes.

First Period.

Practice of the company in the third class.

126. Every man will commence the yearly course of practice in the third class, and be exercised therein at sixteen different distances in four practices.

127. At the conclusion of these practices the columns in the company judging distance practice return will be totaled up, and receive the signature of the captain of the company to verify its correctness, as also of the officer-instructor, who will previously compare it carefully with the " duplicate total points " in his possession.

128. All those men who obtain sixteen points will pass into the second class, the remainder will recommence the practice of the third class.

Second Period.

Practice of the second and third classes.

129. Each company will now be told off into two classes and into sections, and the practices continue in that order. Each class will be exercised at sixteen different distances in four practices.

130. At the conclusion of the practice in the second period, the columns of this period in the company judging distance practice return will be totaled up and signed by the captain and officer-instructor as before.

131. All those men who in the practice of the second class obtain sixteen points will pass into the first class, the remainder will repeat the practice of the second class. The test for passing from the third to the second class will be the same as in the practice of the first period.

Third Period.

Practice of the first, second, and third classes.

132. The company will now be told off into three classes, and into sections as before, and each class exercised at sixteen different distances in four practices.

133. The second class will be composed partly of men who repeat the practice of the second class, and partly of men who have passed out of the third class in the second period.

134. At the conclusion of the practices in the third period, the columns of this period in the company's judging distance practice return will be totaled up and signed as directed for the first and second periods. A final classification will then be made, and the man who, in the practice of the first class, obtains the greatest number of points will obtain the battalion prize as the best judge of distance. Should two or more men obtain the same number of points, the prize will be awarded to that man who has obtained the greatest number of points throughout the whole practice.

INSTRUCTION OF RECRUITS.

135. Every recruit, before he is allowed to join the practice of the battalion, will be put through the foregoing course, with the exception of the *judging distance practice*, under the close superintendence of the officer-instructor and his assistants.

136. In the aiming drill the instructor should at first cause the recruit to aim at a small mark on the wall of the yard or barrack room, and confine his attention to those rules laid down under the head of aiming drill.

137. After the recruit has been well grounded in the various exercises under the head of drill, and before he is allowed to fire ball, he will be practised in *snapping caps and firing blank cartridge* to give him steadiness, and accustom him in some measure to the explosion of the gunpowder and recoil of the piece.

138. Caps will be first used, after which, each recruit will be made to fire two or three rounds singly, then a few rounds in file firing and volleys. Should the instructor experience any difficulty in teaching any of the recruits to aim, or in practice find any man shooting badly, he will distribute three or four caps to them, and, having placed a lighted candle on a table or stand at eight or ten paces in front of the squad, he will then make each man advance in succession to such a distance from the light that, when aiming, the muzzle may be about a yard from it, and after going through the motions of loading and putting on the cap, the man will fire, aiming at the wick of the candle, when, if the aim is properly directed, the candle will be blown out.

139. After all these exercises have been gone through, the soldier will be competent to join the practice of his

D

battalion; but any man who concludes his practice as a recruit after the target practice of his battalion has commenced, will not fire with his battalion until the ensuing year.

140. An index of preliminary drills will be kept by the non-commissioned-officer instructor of the battalion for the recruits, as also a target practice return of form marked F, recording their practice.

PRIZES.

141. Prizes will be awarded at the conclusion of the whole practice, subject to such arrangements as may be hereafter approved by the Field-Marshal Commanding in Chief.

RETURNS, &c.

142. The following forms will be made use of in the different branches of the instruction, viz. :—

Company Returns.

FORM A. A company index return of preliminary drills, in which the men's names are to be entered by sections or squads, the non-commissioned officers heading each section. This return is to be filled in by the company instructor after each drill or parade, for which the captain is responsible.

FORM B. A register of target practice, for each squad or section, to record the practice as it proceeds, the men's names to be entered in the order in which they stand in the company's index and practice returns.

FORM C. A register of judging distance practice, for each squad or section, to record the practice as it proceeds, the men's names to be entered in the order they stand in the company's index and practice returns.

FORMS D and E. Diagrams to record the file and volley firing, and skirmishing practice of each section.

FORM F. A company target practice return, to be filled in as the several practices occur by the company instructor, the men's names to be inserted therein by sections, the non-commissioned officers heading each section. As this return is a record showing the progress of the company in its practice, as well as its efficiency in the use of the rifle, the captain is to be held responsible that it is kept with great care and correctness.

FORM G. A company judging distance practice return, to be filled in, as the several practices occur, by the company instructor, the men's names to be entered therein by sections, the non-commissioned officers heading each section:—the captain will be held responsible that this return also is kept with care and correctness.

FORM H. A monthly return of target and judging distance practice, showing the progress of instruction in the battalion during the month, to be prepared by the officer-instructor; and, after being signed by him, and countersigned by the commanding officer, to be transmitted in duplicate to the school of musketry on or before the fourth of the month following that for which it is made out; one copy will be returned with any remarks it may be found necessary to make. *Battalion Returns.*

FORM I. A battalion target practice return, to be prepared in duplicate by the officer-instructor, and rendered at the autumn half-yearly inspection of the regiment; one copy to be given to the inspecting general officer, and the other transmitted to the school of musketry.

FORM K. A battalion judging distance practice return, to be prepared by the officer-instructor in duplicate, and rendered at the autumn half-yearly inspection of the regiment; one copy to be given to the inspecting general officer, and the other transmitted to the school of musketry.

143. In the event of there being district inspectors these returns will be sent through them.

144. A company return of the form marked F, signed by the officer-instructor, and countersigned by the commanding officer, showing the practice of the recruits during the year, will be transmitted with the annual target and judging distance practice returns.

145. At the conclusion of the annual course of instruction the registers of target and judging distance practice will be handed over to the officer-instructor, who will take charge of them until he receives instructions for their disposal; the practice returns will be retained in the regiment as public records, liable to be called for at any time, of the efficiency of the several companies.

RECAPITULATION of the NUMBER of DRILLS or PRACTICES in the Instruction of Musketry, to be gone through by every Non-commissioned Officer and Soldier of the Battalion annually, and by the Recruits before they join in the Practice of the Battalion.

PRELIMINARY DRILLS.	N. C. Officers and Soldiers.		Recruits.		REMARKS.
	No. of Drills or Practices.	No. of Rounds.	No. of Drills or Practices.	No. of Rounds.	
Theoretical principles -	6	..	According to the discretion of the Officer-Instructor	..	By the Officer-Instructor.
Cleaning arms - -	6	..		..	By the N. C. Officer-Instructor.
Target drill { Aiming drill	6	..		..	By the Officer-Instructor.
Target drill { Position drill	6	..		..	Ditto.
			No. of Percussion Caps.	Blank Cartridges.	
Snapping caps, and blank cartridges - - -	..	..	20	20	Ditto.
Judging distance drill -	12	..	According to the discretion of the Officer-Instructor	..	Ditto.

PRACTICES.				Ball Cartridge.	
Preliminary firing, one round to be fired from a rest at the several distances to 300 yards -	..	..	5	20	By the Officer-Instructor. N.B. These 20 rounds are to be recorded in a register, but not in the company's practice return.
Individual firing { 1st Period - -	5	20	5	20	In the company under the captain of companies, and recruits by the Officer-Instructor.
Individual firing { 2nd Period { 2nd Class	6 }	20	6 }	20	By the Officer-Instructor.
Individual firing { 2nd Period { 3rd Class	5 }		5 }		
Individual firing { 3rd Period { 1st Class	6 }	20	6 }	20	Ditto.
Individual firing { 3rd Period { 2nd Class	6 }		6 }		
Individual firing { 3rd Period { 3rd Class	5 }		5 }		
File-firing and volleys -	1	10	1	10	In the companies by their captains, recruits by the Officer-Instructor.
Skirmishing practice -	2	20	2	20	Ditto.
Judging distance practice { 1st Period -	4	..	..	..	By the Officer-Instructor.
Judging distance practice { 2nd Period -	4	..	..	..	Ditto.
Judging distance practice { 3rd Period -	4	..	..	..	Ditto.
Total -	..	90	..	110	

CIRCULAR MEMORANDUM.

Horse Guards, 1st January 1857.

Gen¹ Nº 249
B. 88.
Ins" of Musk,
9.

THE General Commanding in Chief desires, that the accompanying regulations may be substituted for those contained in the printed book of "INSTRUCTION OF MUSKETRY," dated 1st January 1856.

His Royal Highness further directs it to be notified that, as a general rule, future appointments to the office of Instructor of Musketry, at the head quarters of regiments, will be made from the subalterns; and that, whenever the Officer-Instructor, either of a regiment or depôt battalion, may be absent for upwards of fourteen days, the allowance for the whole period of such absence is to be issued to the officer who may have been appointed the Assistant Instructor.

By Command,
G. A. WETHERALL,
Adjutant General.

APPENDIX.

With a view to establish uniformity in the mode of proceeding with the musketry training, and to ensure the instructions being efficiently carried out in the shortest time, His Royal Highness the General Commanding in Chief desires that the following regulations be strictly observed:

1. On an instructor joining a battalion, and at the commencement of the annual course of instruction, the commanding officer is to hand over to the instructor the full non-commissioned officers of the battalion, by one fourth at a time, to be exercised through a course of preliminary drills, and to be struck off all duty while under instruction.

2. When all the serjeants and corporals have been exercised through the preliminary drills, which can be well effected in one month, they are to be employed to assist in the instruction of their respective companies as herein-after detailed, and the training of the battalion is then to be proceeded with, by companies, in the following manner; the colour serjeant of each being appointed the company-instructor.

3. A company of the right wing, with its officers and non-commissioned officers, is to be struck off all duty, and handed over to the officer instructor, who will first exercise it in the preliminary drills, which can be performed in six days, viz.:

—		Theoretical Principles.*		Target Drill. Aiming Drill.		Target Drill. Position Drill.		Judging Distance Drill.		Cleaning Arms.		Remarks.
		Time occupied.	No. of Drill.	Time occupied.	Distances at which to aim.	Time occupied.	No. of Drill.	Time occupied.	No. of Drill.	Time occupied.	No. of Drill.	
1st Day	A.M.	½ hour	1	¾ hour	Explain the sights and aim at 100 yards.	½ hour	1	¾ hour	1	—	—	When at Aiming Drill, those men not actually engaged in aiming to be practised in "Position Drill," with the sight or elevation for the actual distance, so that the time may be profitably employed.
	P.M.	—	—	½ hour	150 yards.	—	—	¾ hour	1	½ hour	1	
2nd Day	A.M.	½ hour	1	¾ hour	200 & 250 "	½ hour	1	¾ hour	1	—	—	
	P.M.	—	—	½ hour	300 "	—	—	¾ hour	1	½ hour	1	
3rd Day	A.M.	½ hour	1	¾ hour	350 & 400 "	½ hour	1	¾ hour	1	—	—	
	P.M.	—	—	½ hour	450 "	—	—	¾ hour	1	½ hour	1	
4th Day	A.M.	½ hour	1	¾ hour	500 & 550 "	½ hour	1	¾ hour	1	—	—	
	P.M.	—	—	½ hour	600 "	—	—	¾ hour	1	½ hour	1	
5th Day	A.M.	½ hour	1	¾ hour	650 & 700 "	½ hour	1	¾ hour	1	—	—	
	P.M.	—	—	½ hour	750 "	—	—	¾ hour	1	½ hour	1	
6th Day	A.M.	½ hour	1	¾ hour	800 & 850 "	½ hour	1	¾ hour	1	—	—	
	P.M.	—	—	½ hour	900 "	—	—	¾ hour	1	½ hour	1	
TOTAL -	-	—	6	—	17 Distances in 6 Drills.	—	6	—	12	—	6	

* In addition to the time here specified, wet days are to be taken advantage of, to ascertain the proficiency of the men in this subject.

4. After this company has gone through the preliminary drills, it is to proceed immediately to target and judging distance practice, when in six days more it is to be exercised through the 1st and 2nd periods of individual firing,—each man expend-

ing 40 rounds of ball ammunition,—as also, through the 1st and 2nd periods of judging distance practice, when it will return to duty.

5. When the company referred to in paragraph 4 proceeds to target practice, &c., a company of the left wing, with its officers and non-commissioned officers, is to be struck off duty, exercised in the preliminary drills as per preceding table, so as to be prepared to take the place of the company at practice by the time it has completed the 2nd period; and when it has done so, another company of the right wing is to replace it; and the instruction is to be continued, by companies of alternate wings, until the whole battalion has been similarly exercised, which, in moderate weather, can be accomplished in two months, as shown by the following statement, by which it will appear that a company need only be kept off duty for a fortnight (weather permitting), and that one-fourth of the battalion during the said two months will be employed in training to the use of their arms, viz., a company of one wing at preliminary instruction, and a company of the other wing at practice, after which period, only one company or one-eighth of the battalion will be required at a time for further practice until the annual course is completed.

The annual course of Instruction to commence with the Training of the non-commissioned officers of the Battalion on the 15th March, so that on the 15th April it will commence in the Companies.

	Preliminary Drills.		Target and Judging Distance Practices, 1st and 2nd Periods.	
	From	To	From	To
A Company, Right Wing	15th April	20th April	22nd April	27th April.
A Company, Left Wing	22nd „	27th „	29th „	4th May.
A Company, Right Wing	29th „	4th May	6th May	11th „
A Company, Left Wing	6th May	11th „	13th „	18th „
A Company, Right Wing	13th „	18th „	20th „	25th „
A Company, Left Wing	20th „	25th „	27th „	1st June.
A Company, Right Wing	27th „	1st June	3rd June	8th „
A Company, Left Wing	3rd June	8th „	10th „	15th „

6. When every company of the battalion has been exercised to the end of the 2nd period of target and judging distance practices, a company, with its officers and non-commissioned-officers, alternately by wings, is again to be struck off duty,—exercised in the 3rd period of individual firing, and judging distance practice, in file and volley firing, and two skirmishing practices,—all of which may be executed in a week, so that in two months the whole battalion will have been instructed.

7. By the foregoing arrangement, a battalion of eight companies should complete in an efficient manner its annual course of musketry drill and practice as prescribed in the "*Instruction of Musketry*": in five months, each man firing 90 rounds of ball ammunition.

8. There are very few stations at which the foregoing directions will not apply; but where the duties may be so severe as to render it impossible, even for the short period of two months, to strike one fourth of a battalion off duty, then only one company, with its officers and non-commissioned-officers, alternately by wings, is to be handed over at a time for instruction, and every company is to be exercised to the end of the 2nd period before any further advance is made in the yearly course of instruction, which is then to be proceeded with as laid down in paragraph 6.

9. In depôt battalions, after the non-commissioned-officers have been instructed as directed in paragraph 1., one company at a time from each depôt, with its officers and non-commissioned officers, is to be struck off duty, and handed over to the instructor, to be exercised as before directed.

10. Provisional battalions such as that at Chatham are to furnish for musketry instruction four officers, four serjeants, and 100 rank and file, with four company instructors, to be employed at practice, and four officers, four serjeants, and 100 rank and file with four company-instructors, to be engaged with the preliminary drills; the former can be exercised to the end of the 2nd period (each man firing 40 rounds), in a week, by which time the latter will have completed the preliminary drills, and be prepared to replace them, when an equal number is again to be furnished for drill; this relief to continue until all the effectives present have been exercised to the end of the 2nd period of the target and judging distance practices,— by which means the instruction will proceed uniformly throughout the battalion. When this is accomplished, similar parties are to be handed over to be exercised in the 3rd period; each company or depôt furnishing men in proportion to its effective strength present, to make up the number required to be under instruction. The file and volley firing and skirmishing practices are also to be conducted in the same manner.

11. The annual course of musketry instruction is invariably to commence in every battalion in Great Britain, Ireland, and the Channel Islands on the 15th March, in order that it may be finished by the autumn half-yearly inspection; when the annual target and judging distance practice returns are to be rendered to the general officer, and copies of them are to accompany his confidential report.

12. At Malta, Gibraltar, in the West Indies, and at other stations abroad where the heat is such as to prevent the musketry exercises proceeding during the summer months, the annual course of instruction is to commence on the 15th of September. In such cases the annual target and judging distance practice returns are to be rendered on the 15th April.

13. Squads of recruits are to be handed over entirely to the officer-instructor for musketry training, in the first instance for three weeks, in which time it is expected that, with few expections, every man will be well grounded in the preliminary exercises, and have practised to the end of the 2nd period of individual firing, in addition to the preliminary practice. The squads are to be subsequently handed over to the instructor for one week more, to complete the 3rd period of individual firing, the file and volley firing, and the two skirmishing practices.

14. To expedite the musketry training of the recruits, one squad is to be under preliminary instruction, and another squad engaged at practice at the same time. When the number of recruits in a battalion exceeds 60, each squad handed over for musketry instruction is to consist of 20; if less than 60, each squad is to consist of 10.

15. As it is of importance that the recruit should be able to handle his arms with freedom before he is sent to the officer-instructor, he is to be under the instruction of the adjutant for a month after he joins the battalion, in which time he must be instructed in all the motions of loading standing and kneeling.

16. When drafts are sent from the depôt to the service companies, a company's practice return, prepared and signed by the officer-instructor, and countersigned by the commanding officer, showing the proficiency of the men, and the stages to which they have been exercised in the annual course of instruction, is to be transmitted with them.

17. All men who, from any cause whatever, have been absent from the preliminary drills during the period the battalion was proceeding with the annual course of training, and all men who remain in the third class at the final classification of the target practice, are to be exercised during the winter months. In regiments abroad, such men are to be exercised early in the morning and in the evening during the summer months. The results of the practices of these men are to be recorded in the recruits' practice return after the recruits, and distinguished as "absentees," "3rd class shots."

18. The foregoing paragraph is not to be interpreted as sanctioning the absence of men when the company to which they belong is proceeding with its musketry training; on the contrary, every soldier is to be present; the only exceptions being, sick in hospital, men absent from head-quarters, in prison, those for whom arms are not supplied, and lads incapable of bearing arms.

19. By the following simple arrangement when judging distance by classes in the 2nd and 3rd periods, and when the ground is tolerably level, much time may be saved, the practice made more effective by precluding the possibility of a clue being obtained to the correct distance, and a greater number of men exercised in classes together. Thus :—the instructor will send forward a party of one of the classes (say the third)—with a non-commissioned-officer instructor—as "points" from which the several classes will judge their distance, the party as "points" at the same time estimating the distance from the class to which it belongs; the non-commissioned-officer instructor in charge of the "points," and the commander of each class, being furnished by the instructor on the practice ground with a memorandum showing the distance from the end of the chain at which the party as "points" are to stand; the correct distance at which each class is situated from the "points" judged from, will be ascertained by deducting the distance at which the "points" are stationed, from that at which the class is found to be from the end of the chain. Care must, however, be taken that the several classes so place themselves as not to prevent those in rear of them from seeing the "points."

20. When the ground is so irregular or hilly that the judging distance practice cannot be conducted according to form, the instructor should exercise his men by obtaining the distance of any conspicuous objects by triangulation, recording the answers in the usual way.

21. Captains or officers commanding companies are held responsible that the following returns are carefully and correctly kept, as they are the records showing the progress and efficiency or otherwise of each soldier of the company in the use of his rifle: they are simple, and easy to keep, if filled in as the several practices occur.
Index Return of Preliminary Drills.
Target Practice Return.
Judging Distance Practice Return.

22. These returns, whenever required, are to be sent to the officer-instructor, who is to bring to the commanding officer's notice any neglect he may observe therein.

23. To establish uniformity in the mode of preparing the monthly progress returns of regiments, depôt-battalions, &c., the following rules are to be strictly attended to:—

"*Strength, &c.*"—Under this head are to be shown the effectives of each company on the last day of the month, minus recruits, as explained in the following paragraph, showing in the margin of the return to the right of the column of remarks, by ranks and companies, the number of men absent from head quarters, recruiting, in hospital, in prison, &c. Servants and men in regimental employ,—all of whom are to be trained as the other soldiers annually,— are not to be included among the absentees.

"*Recruits.*"—Under this head are to be shown all men effective on the last day of the month, who, when the annual course of instruction commenced, had not concluded the practices prescribed in paragraph 135, page 49, of the Book of Instruction, as also all men who have joined the battalion as recruits after that date, such men, however, being considered recruits only with reference to their musketry training. When a battalion commences for the first time the course of instruction, the men who on the last day of the month in which the course commenced were not dismissed drill, and all men who have joined after that date, are to be classed as recruits.

"*Number under instruction, &c.*"— Here is to be shown the number of effectives who have been under instruction since the annual course commenced, and the practices in which exercised. Should a company or squad of recruits not have concluded the "period" or practice in which it is exercising on the last day of the month, the words "in progress" are to be inserted under the said practice opposite the company or squad, to show the practice in which it is engaged. Under this head four lines are circumflexed for recruits. In the first and second lines is to be shown the number of the effective recruits who have been under instruction, and the stages to which exercised; those farthest advanced in their training to appear in the top line. In the third line is to be shown the number in progress with the target practice on the last day of the month, and in the fourth line the number in progress with the preliminary drills.

"*Number in each class, &c.*"—In this table the classification of the men at the conclusion of the periods of practice in which they have exercised is to be shown. Should a man, having commenced a period, become a casual, and not be able to conclude it, he is to be classified according to the merit of his performances therein. The numbers classified should agree with the numbers shown in the body of the return to have been exercised in the periods to which said classification refers,—if they do not, then the cause of discrepancy is to be explained in as brief a manner as possible in the column of remarks. As the number of recruits with their practices is shown distinctly from the other soldiers of the battalion, their classification is also to be recorded separately in brackets on the right of the number referring to the soldiers of the battalion,— thus :—44 (12). Should any man, recruit or otherwise, become a casual when proceeding with the preliminary drills, he is to be considered as not having been exercised therein.

24. The instruction of musketry, like all other duties, being under the supervision of commanding officers, who are accountable for the accuracy of all returns, &c., the General Commanding-in-Chief directs that they will give every information and explanation which may be required thereon by the Inspector General at Hythe, who is responsible that the system is conducted efficiently and according to regulation.

G. A. WETHERALL,
A. G.

List of Articles allowed for the Instructor of Musketry, and to be obtained, on application to the Barrack Master, at the several Stations.

	Articles required.	No. of each required.	Remarks.
For each Barrack.	Iron targets according to extent of range.		
	Flags { Red - - - - 6 feet square	1	
	Red - - - - 2½ „	3	
	Red and white - - 2½ „	2	
	Dark blue - - - 2½ „	2	
	White - - - - 2½ „	2	
	Poles, lance 10 feet long - - - -	10	
	900 yards length of Gunter's chain or cord, labelled every 5 yards, and numbered from 1 to 900, divided into 18 equal parts - - -	1	
	Pins of stout wire, 12 inches long - - -	18	
	Tripod rests with rings - - - -	12	
	Sand bags (bushel) - - - - -	12	
	Large brushes for colouring targets (lbs. brushes) -	2	
	Small „ „ „ sash tools -	2	
	Whiting - - - - - - -		*Sufficient for the annual supply of a battalion 1,000 or 1,200 strong, to be demanded from barrack-master in small quantities.
	Lamp black - - - - - - -	*	
	Glue to make size - - - - - -		
For each Lecture Room.	Black board 6 feet × 4 feet and easels - -	1	
	Cap with cross wires to fit muzzle of rifle musket, pattern 1853 - - - - - -	1	
	Wooden plug with hole through the centre, to fit into breech end of rifle musket, pattern 1853 -	1	
	Wooden model with suspended wires to illustrate the necessity of holding the sight upright when taking aim - - - - - - -	1	
	Common flat ruler - - - - 3 feet long	1	
	Rifle musket barrel, pattern 1853 - - -	1	
	Waster locks with cocks complete, swivel pattern -	6	
	Improved turnkey with cramp, worm, &c., attached	6	
	Implements for making cartridges - - set	1	
	Set to consist of— 5 Tin measures, containing each 2¼ drams. 5 Tin funnels with long narrow spouts. 12 Mandrels of hard wood for cartridges for rifle musket, pattern 1853. 12 Formers „ for „ do. do. 1 Set of tin patterns, showing shape of paper for cartridges. 1 Iron straight edge, 1 large knife.		
	20 Quires of white paper for inner and outer envelope of cartridge - - - -	} Required Annually.	
	6 Quires of cartridge paper for cylinder of cartridge		
	½ Bushel of fine sand - - - - -		
	50 Bullets for rifle musket 1853.		
	12 Files to retain documents (common).		
	Sponge - - - - - - -	2 oz.	
	Chalk, common - - - - - - lb.	1	
	Circular ruling pens to describe circles or diagrams	2	
	Compasses with holder to contain chalk - -	1	

PLATE 1

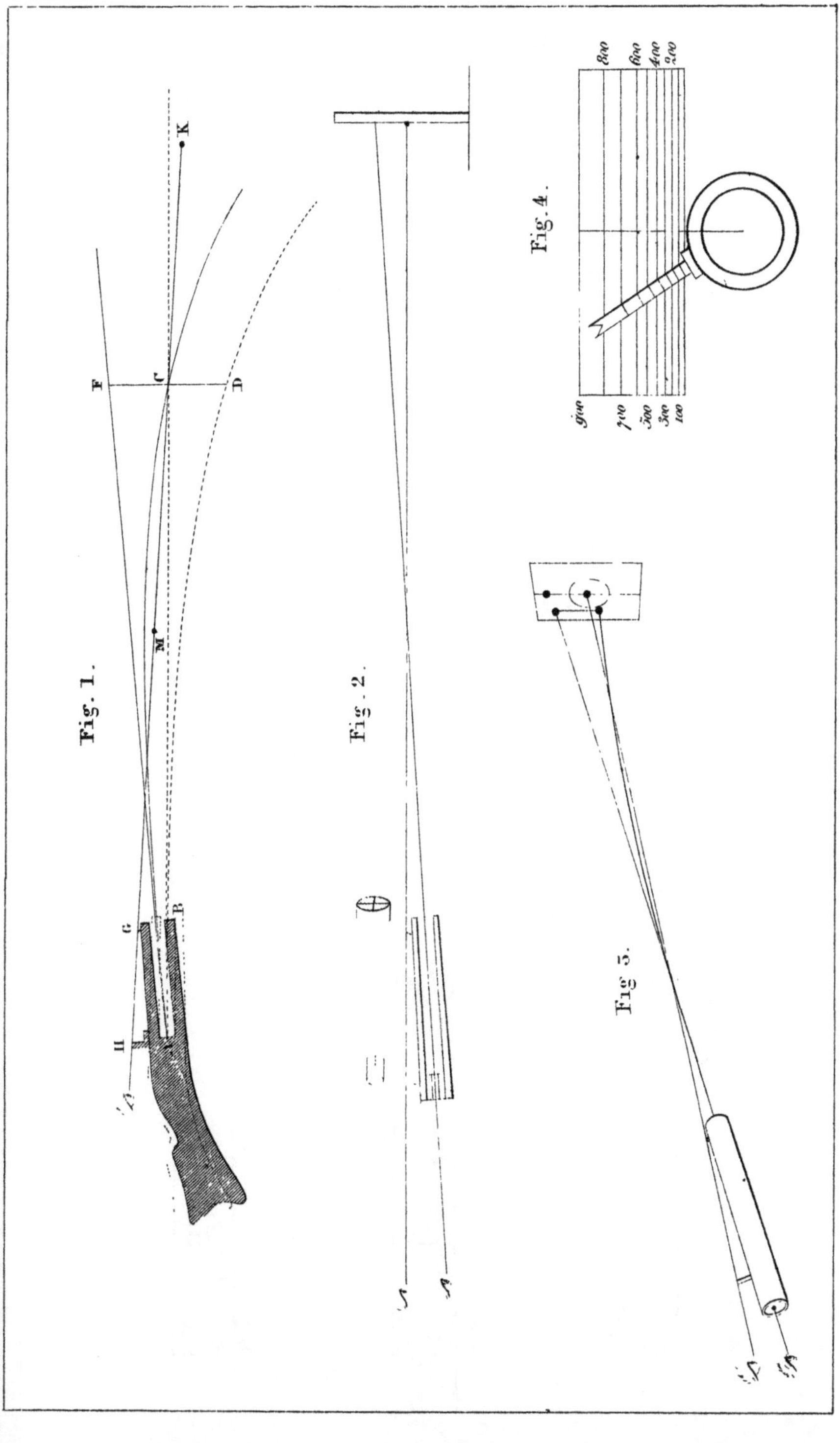

Fig. 1.

Fig. 2.

Fig. 3.

Fig. 4.

Day & Son, Lith^{rs} to the Queen

Fig. 5.

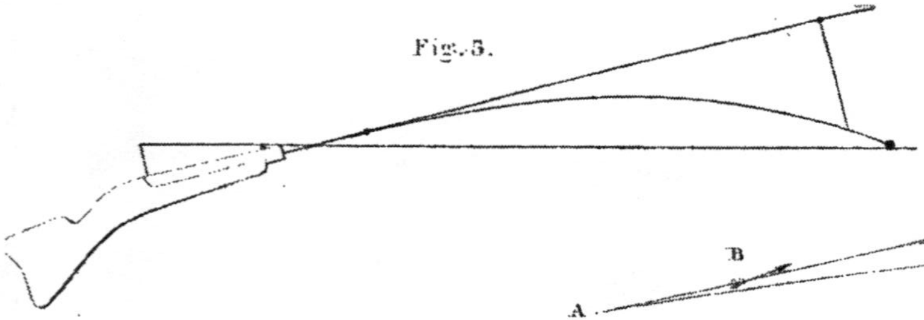

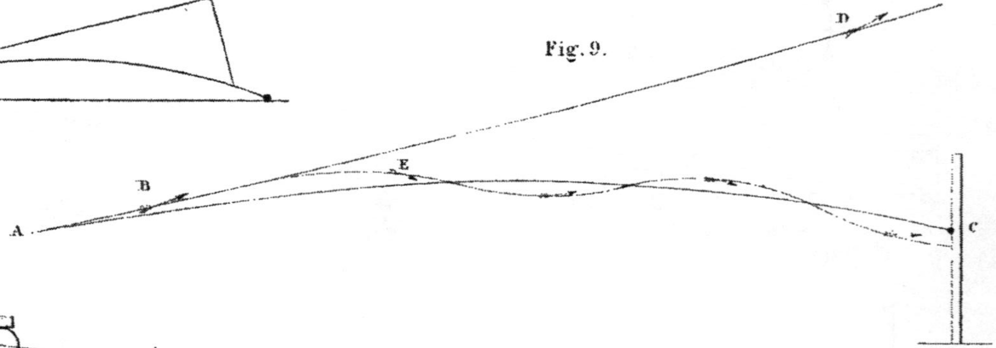

Fig. 6.

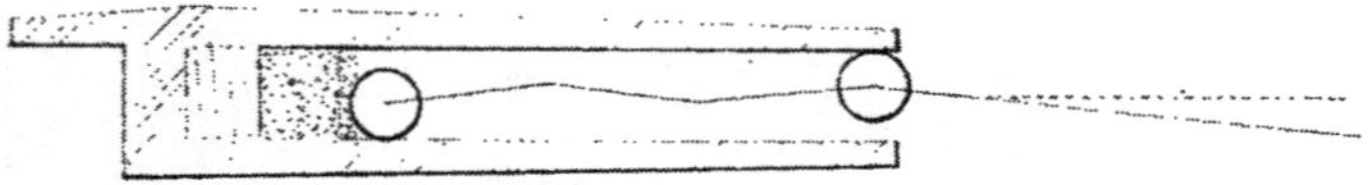

Fig. 7.

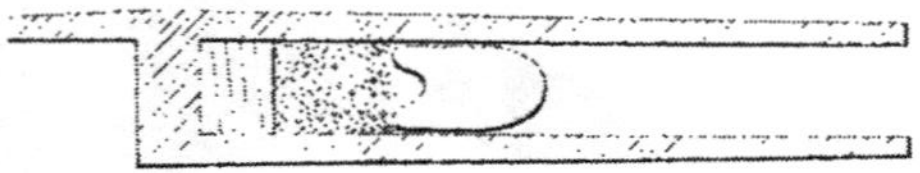

Fig. 9.

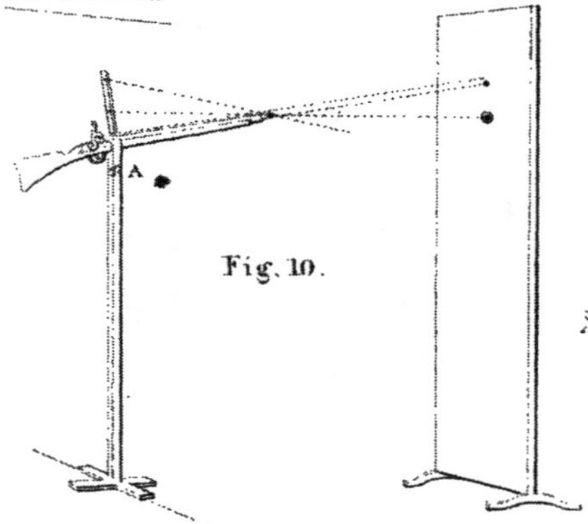

Fig. 8.

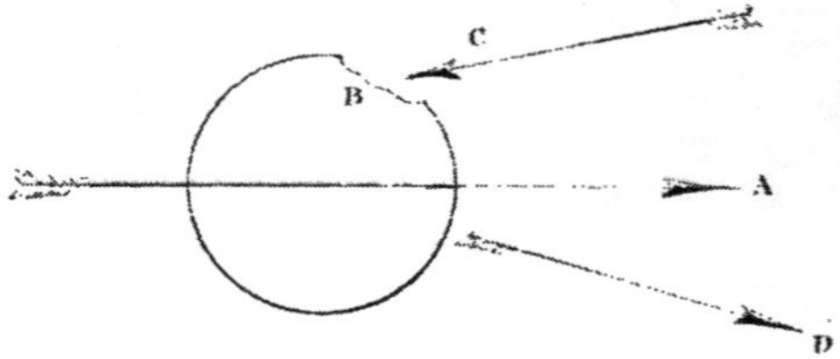

Fig. 10.

4 Feet

8 Feet

Fig. 11.

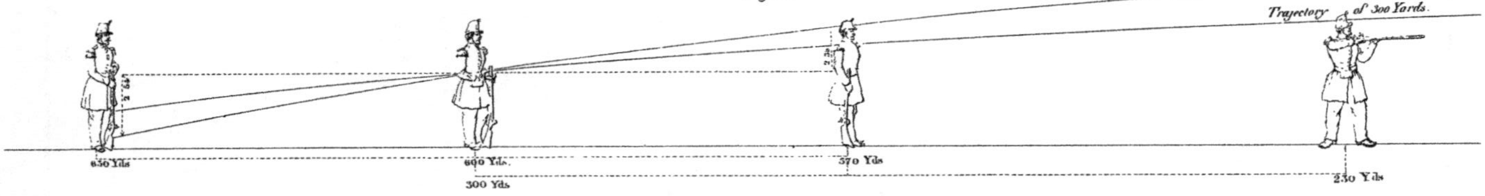
Trajectory of 600 Yards.
Trajectory of 300 Yards.
650 Yds
600 Yds.
370 Yds
300 Yds
230 Yds

Fig. 12.

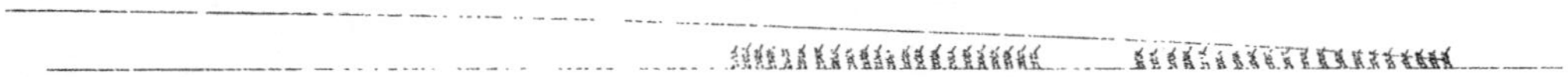

Day & Son, Lith^rs to the Queen

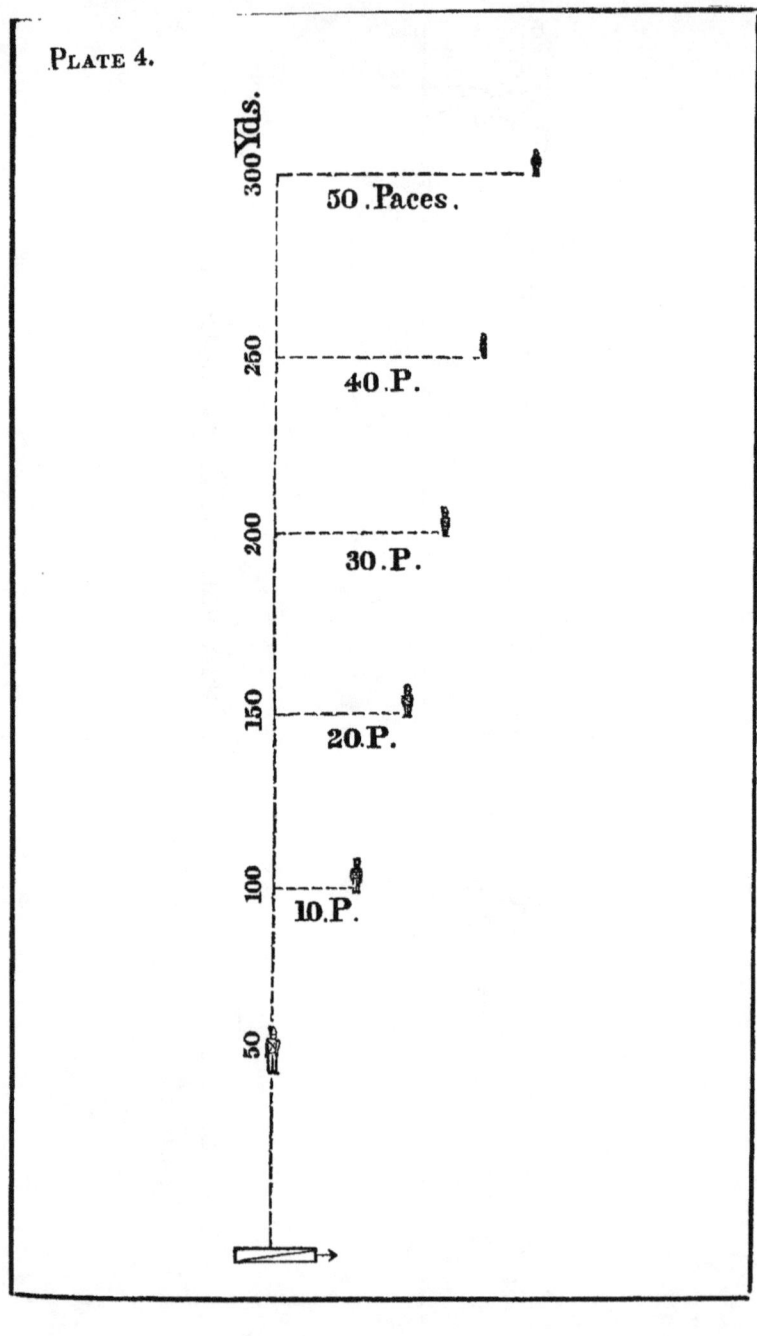

PLATE 5

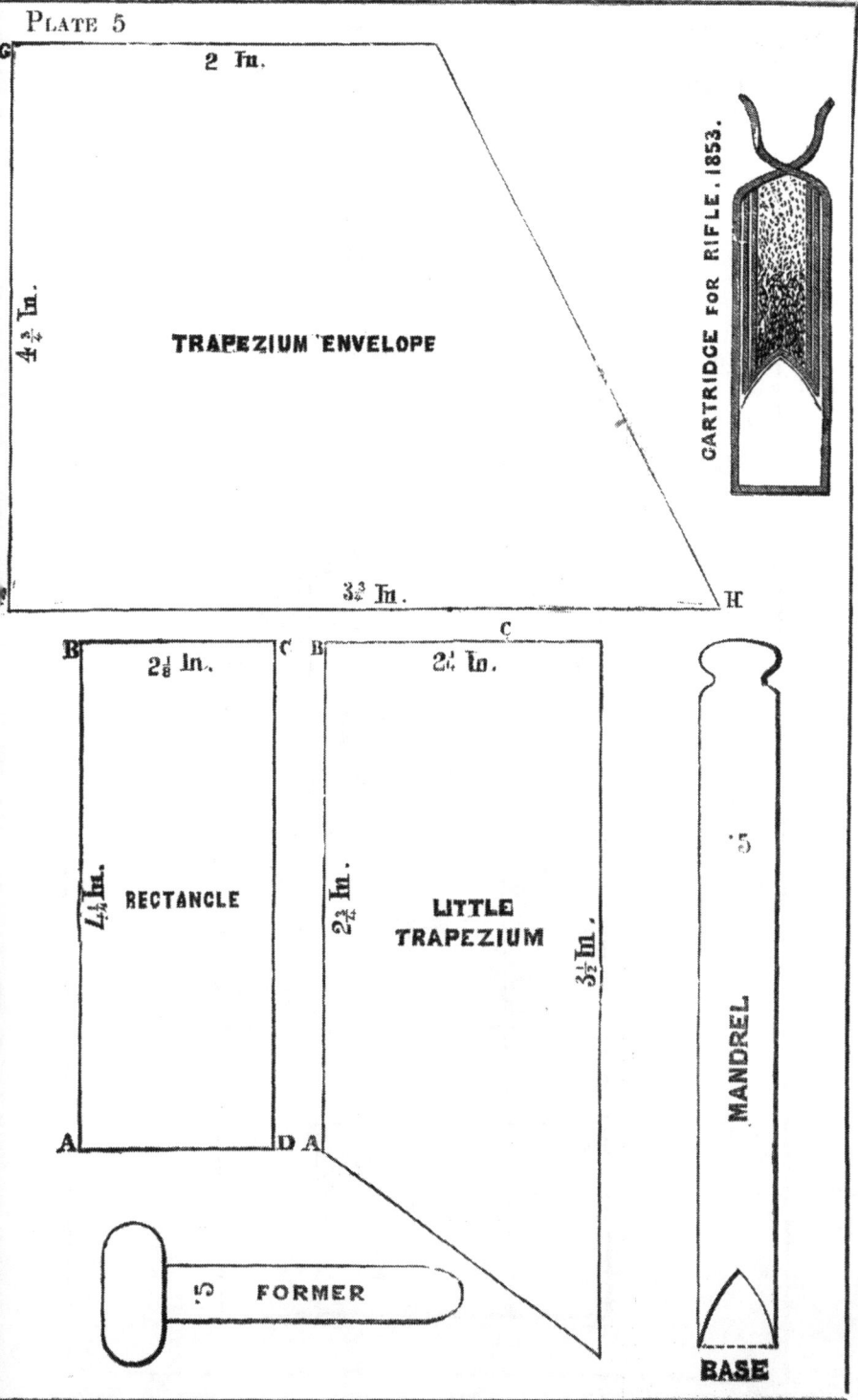

PLATE 6.

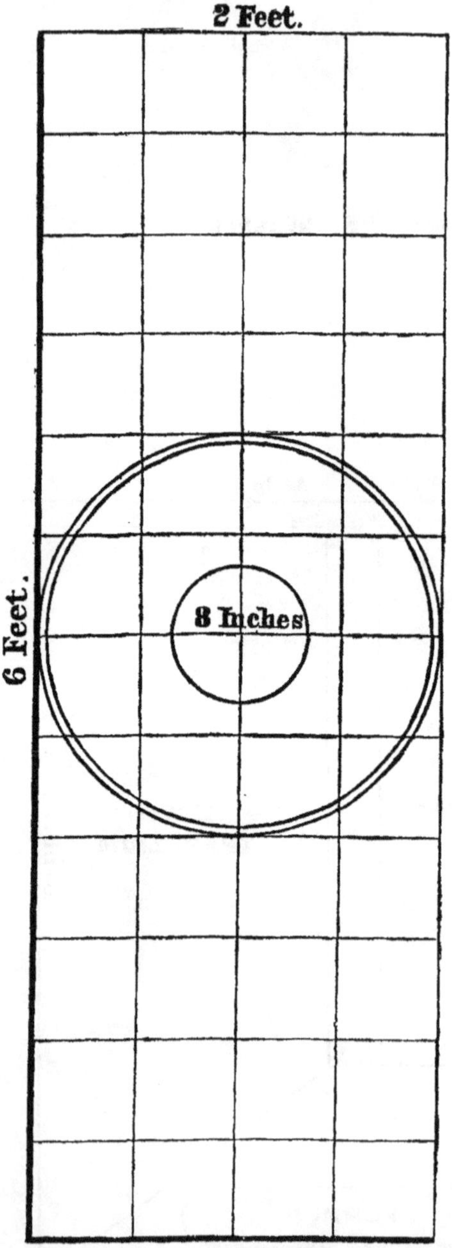

2 Feet.

6 Feet.

8 Inches

PLATE 7.

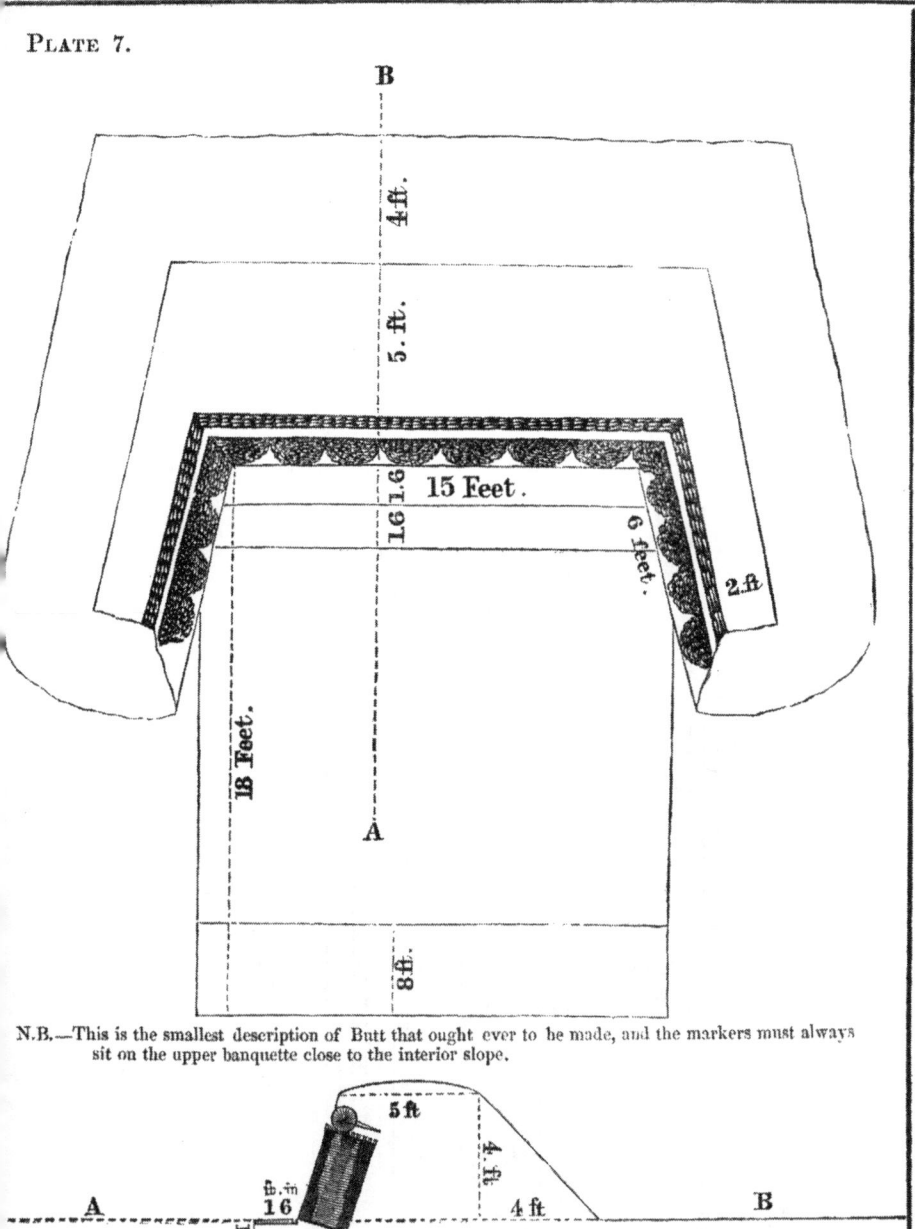

N.B.—This is the smallest description of Butt that ought ever to be made, and the markers must always sit on the upper banquette close to the interior slope.

Section on the line A.B.

PLATE 8.

A. No.______________ Company ______________ Battalion ______________ Regiment.

INDEX for PRELIMINARY DRILLS performed by Capt. ______________'s Company during the year 18___. Date ______________ 18___

No.	RANK AND NAMES. Recruits not included.	Theoretical Principles.						Cleaning of Arms.						Target Drill.												Judging Distance Drill.															REMARKS, Showing the Cause of Absence from any of the Exercises here nam
														Aiming Drill.						Position Drill.																					
		1	2	3	4	5	6	1	2	3	4	5	6	1	2	3	4	5	6	1	2	3	4	5	6	1	2	3	4	5	6	7	8	9	10	11	12	13	14	15	
1																																									
2																																									
3																																									
4																																									
5																																									
6																																									
7																																									
8																																									
9																																									
10																																									
11																																									
12																																									
13																																									
14																																									
15																																									
16																																									
17																																									
18																																									
19																																									
20																																									
21																																									
22																																									
23																																									
24																																									
25																																									
26																																									
27																																									
28																																									
29																																									
30																																									
31																																									
32																																									
33																																									
34																																									
35																																									
36																																									
37																																									
38																																									
39																																									
40																																									
41																																									
42																																									
43																																									
44																																									
45																																									
46																																									
47																																									
48																																									
49																																									
50																																									
51																																									
52																																									
53																																									
54																																									
55																																									
56																																									
57																																									
58																																									
59																																									
60																																									
61																																									
62																																									
63																																									
64																																									
65																																									
66																																									
67																																									
68																																									
69																																									
70																																									
71																																									
72																																									
73																																									
74																																									
75																																									
76																																									
77																																									
78																																									
79																																									
80																																									
81																																									
82																																									
83																																									
84																																									
85																																									
86																																									
87																																									
88																																									
89																																									
90																																									
91																																									
92																																									
93																																									
94																																									
95																																									
96																																									
97																																									

*Officer-Instructor.* _*Captain Commanding Com*_

PLATE 9.
B. FORM 929.

Register of Target Practice.

1st Period.   3rd Class.   2nd Company.   5th Regiment.
Targets—Two.   1st Section.   Distance—Three hundred yards.   Date—August 16, 1855.

1st Sect. 2nd Compy.
1st Period, 3rd Class.
Distance, 300 yards.
Date 16?/53.
T. P. K.

	RANK AND NAMES.	1	2	3	4	5	6	7	8	Total Points.	Casuals.	Casuals.	REMARKS.	Duplicate Total Points.
1	Serjeant Nobes	1	1	0	3					5				5
2	Corporal Hills	3	0	1	1					5				5
3	Private Andrews	0	1	0	3						4	4		4
4	„ Burrage	0	0	3	1									4
5	„ Gardner	0	3	2	1					6				6
6	„ Pullen	0	0	1	2						3	3		3
7	„ Mitchell	1	2	2	0					5				5
8	„ Talbot	1	1	1	0					3				3
9	„ Palmer	3	2	0	1					6				6
10	„ Barker	0	0	0	3					3		3		3
11	„ Smith	1	1	0	1					3				3
12	„ Bates	0	1	2	1					4				4
13	„ Cross	3	0	0	3						6			6
14	„ Butler	1	0	0	3					4				4
15	„ Tyler	0	2	0	2					4				4
16	„ Cherry	2	2	0	1							4		4
17	„ Crabb	3	1	1	2					7				7
18	„ Burgess	1	0	1	3					5				5
19	„ Martin	0	1	0	0					1				1
20	„ Poynter	3	1	0	1					5				5
21														
22														
23														
24														
25														
26														
27														
	Total									83	13	11		87
	Total divided by the number of Men									4·5	4·33	3·66		4·35

(Signatures)   J. FORT,   T. HILLS,   C. LUCAS,   C. L.
             Marker.   Company Instructor.   Officer-Instructor.

D 9

C. No. 10.—FORM 930. Register of Judging Distance Practice.

2nd Practice. 2nd Period. 2nd Class. 1st Section. 2nd Company.
State of Atmosphere—Cloudy. Object—Two Men. Date—8 September 1855.

1st Sect, 2nd Company. 2nd Practice. 2nd Period. 2nd Class. Date 8/9/55. J.D.R.

RANK AND NAMES.	1 Correct Dist. 320. Answers.	1 Points.	2 Correct Dist. 395. Answers.	2 Points.	3 Correct Dist. 460. Answers.	3 Points.	4 Correct Dist. 535. Answers.	4 Points.	5 Correct Dist. 460. Answers.	5 Points.	6 Correct Dist. 290. Answers.	6 Points.	7 Correct Dist. 370. Answers.	7 Points.	8 Correct Dist. 300. Answers.	8 Points.	9 Correct Dist. Answers.	9 Points.	Total No. of Answers.	Total No. of Points.	Duplicate Total Points.
1 Private Wilcock	310	0	355	0	400	2	510	1	445	2	280	2	340	1	460	0			4	5	5
2 " Bates																					
3 " Aldridge	310	2	385	2	450	2	550	2											4	8	8
4 " Godmark	310	2	390	2	455	2	505	1											4	7	7
5 " Harnman	310	2	395	2	400	0	540	2											4	6	6
6 " Pointer									480	2	280	2	400	1	480	2			4	8	8
7 " Martin	325	2	400	2	435	1	525	2											4	7	7
8 " Beer	305	2	425	1	450	2	520	2											4	7	7
9 " Burrage	305	1	445	2	445	2	510	1											4	5	5
10 " Hills	325	2	430	0	450	2	550	2											4	6	6
11 " Lennox									470	2	285	3	360	2	445	2			4	9	9
12 " Crabb	305	2	410	2	450	2	515	2											4	8	8
13 " Cherry	305	2	390	2	445	2	510	1											4	7	7
14 " Hake	335	2	400	2	440	2	505	1											4	7	7
15 " Plum	300	2	365	1	445	2	505	1											4	6	6
16 " Snape	320	2	305	0	420	0	580	2											4	4	4
17 " Owens	310	2	395	2	415	0	510	1											4	5	5
18 " Lloyd	320	2	405	2	410	0	525	2											4	6	6
19 " Stevens	325	2	480	0	450	2	550	2											4	6	6
20 " Cross	375	0	460	0	500	0	650	0											4	0	0
21																					
22																					
23																					
24																					
25																					
26																					
27																					
Total																				122	122
Total divided by the number of Men																				6·10	6·10

D   Register of the Practice.   Mullingar, 21.st August, 1855

1st Section .. 4th Company. .. 1st Battalion ...... 97th Regiment..

Distance	300 Yards
Order of Firing	File and Volley
Men	20
Rounds	Each 10 (5 in File firing and 5 in Volley)

Wind. Strong

Atmosphere. Cloudy

Bull's Eyes	6
Centres	25
Outers	55
Lowers	36
Uppers	50
Right	42
Left	44
Hits	86
Misses	114
Rounds	200

Total Number of points   123

Points divided by number of Men   6.15

Best Shots

General Remarks

J. *Hatch*, .. Marker

T. *Money*, Cam.gl Inst.r

C. *Lucas*, .. Officer Instr.

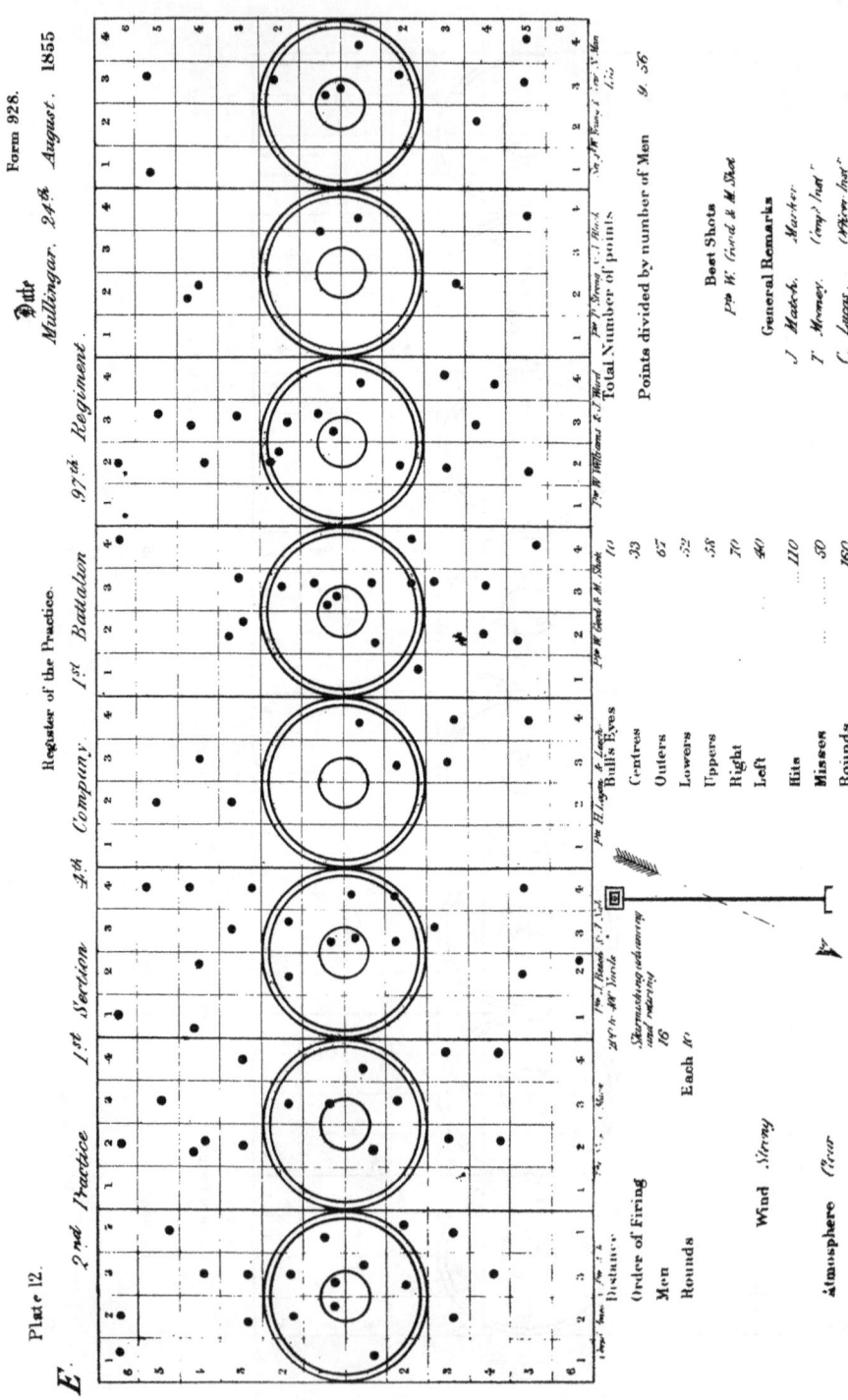

Commenced 23rd March 1855. Concluded 10th August 1855.

[Str]ength of the Company, exclusive of Recruits, as under.	3rd Class. 1st Period. Practice of Company in 3rd Class.					Total Points.	3rd Class. 2nd Period. Practices.					Total Points.	3rd Class. 3rd Period. Practices.					Total Points.	2nd Class. 2nd Period. Practices.						Total Points.	2nd Class. 3rd Period. Practices.						Total Points.	1st Class. 3rd Period. Practices.						Total Points.	Final Classification. Class. 3rd.	2nd.	1st.	File firing and Volleys. Rounds 10. Distance 300 yards. Men present.	Skirmishing. Practice 1. Rounds 10. Distance 160 to 400 yards. Men present.	Skirmishing. Practice 2. Rounds 10.	Remarks.
Practices	1	2	3	4	5		1	2	3	4	5		1	2	3	4	5		1	2	3	4	5	6		1	2	3	4	5	6		1	2	3	4	5	6								
Rounds fired	4	4	4	4	4		4	4	4	4	4		4	4	4	4	4		3	3	3	3	3	3		3	3	3	3	3	3		3	3	3	3	3	3								
Rank and Names. Distances	100	150	200	250	300		100	150	200	250	300		100	150	200	250	300		350	400	450	500	550	600		350	400	450	500	550	600		650	700	750	800	850	900								
ant G. Rushton	7	4	2	3	0	16													0	0	1	0	0	4	5	0	1	0	2	3	1	7								.	1	.	1	1	1	
W. Burnett	2	1	2	1	1	7	4	1	6	0	0	11	0	3	0	1	6	6																						1	.	.	1	1	1	
J. Hobson	1	0	2	1	0	6	4	2	2	0	0	8	4	3	2	1	3	13																						.	1	.	1	1	1	
G. Morris	5	4	3	2	2	16													3	4	1	0	1	1	10								2	0	0	0	1	0	3	.	.	1	1	1	1	
W. Nobes	8	3	1	1	2	15													0	1	2	0	0	0	3	0	2	0	2	0	1	5								.	1	.	1	1	1	
ral T. Ryling	4	3	8	1	0	16	.	.	.	.	.	.	.	.	.	.	.	.																						1	.	.	.	.	.	Hospital.
G. Greeves	4	1	2	1	2	10	.	.	.	.	.	.	.	.	.	.	.	.																						1	.	.	.	.	.	Hospital.
D. M'Beth	2	4	8	3	3	20													5	2	0	1	4	1	13	2	0	0	0	0	1	3								.	1	.	1	1	1	
M. Tuck	6	10	0	0	1	17													1	1	1	4	0	2	9	0	0	1	4	0	2	7								.	1	.	1	1	1	
T. Woods	4	4	1	4	2	15													3	1	1	3	2	3	13								3	0	1	0	0	0	4	.	.	1	1	1	1	
te W. Ackroyd	2	2	2	0	2	8	2	6	0	0	1	9	3	4	1	3	2	13																						.	1	.	1	1	1	
G. Aldridge	2	1	0	2	1	6	2	8	5	2	1	18														1	1	0	1	0	1	4								.	1	.	1	1	1	
H. Andrews	5	1	2	1	2	11	4	3	5	1	1	14														0	1	3	1	2	0	7								.	1	.	1	1	1	
J. Ashton	2	1	3	0	1	7	4	1	3	4	0	12	0	0	0	1	1	2																						1	.	.	1	1	1	
J. Barker	1	0	0	1	1	3	4	6	0	2	0	12	1	1	1	4	1	8																						1	.	.	1	1	1	
R. Bovan	4	3	3	1	2	13	0	4	4	5	1	14														2	1	2	0	0	0	5								.	1	.	1	1	1	
H. Beer	7	5	2	2	1	17													1	0	1	1	1	4	8	2	0	1	2	2	1	8								.	1	.	1	1	1	
T. Bennett	2	5	3	0	4	14													1	1	2	1	1	3	9	1	1	3	3	1	3	10								.	.	1	1	1	1	
J. Burgess	3	2	2	4	1	12	5	0	0	2	1	8	2	1	2	0	0	5																						1	.	.	1	1	1	
G. Bunage	3	7	3	0	0	13													0	3	0	0	1	1	5	0	0	1	1	0	0	2								.	1	.	1	1	1	
W. Rounds	5	0	1	5	3	14													2	4	3	0	2	4	15								2	1	1	4	0	0	8	.	.	1	1	1	1	
J. Bryan	4	2	2	1	2	11	3	6	3	1	3	13														0	0	3	0	0	1	4								.	1	.	1	1	1	
H. Carpenter	3	4	4	2	1	14													1	3	1	1	0	0	6	5	0	1	1	0	4	11								.	.	1	1	1	1	
W. Castle	4	1	6	5	1	17													2	1	0	3	0	3	9	1	1	1	3	3	2	11								.	.	1	1	1	1	
J. Cherry	8	5	5	2	4	24													0	4	1	2	4	3	10								3	0	0	0	0	0	3	.	.	1	1	1	1	
H. Chivers	2	3	0	3	0	8	3	1	3	2	2	11	3	0	2	3	0	8																						1	.	.	1	1	1	
H. Church	4	0	0	2	1	7	5	2	0	2	1	10	4	4	0	3	2	13																						.	1	.	1	1	1	
J. Crabb	4	0	2	1	0	7	2	2	1	1	4	10	1	3	0	1	1	6																						1	.	.	1	1	1	
G. Cracknell	3	2	3	1	2	11	2	1	1	3	0	7	2	3	2	1	2	10																						1	.	.	1	1	.	Absent.
M. Conroy	4	2	2	0	2	10	2	3	0	1	1	7	3	3	7	4	1	18																						.	1	.	1	1	1	
R. Clarkson	5	1	0	1	0	7	4	3	0	2	1	10	0	1	4	0	2	7																						1	.	.	1	1	1	
R. Cross	4	5	1	1	0	11	5	3	2	0	1	11	1	1	0	0	1	3																						1	.	.	1	1	1	
J. Dant	3	1	1	3	2	10	2	4	1	2	1	10	6	3	1	2	3	15																						.	1	.	1	1	1	
G. Drummond	4	1	3	4	1	13													2	0	3	1	0	0	6	1	2	0	0	0	1	4								.	1	.	1	1	1	
Thos. Drysmond	5	3	3	3	3	17													0	2	2	1	1	2	8	1	4	0	1	0	0	6								.	1	.	1	1	1	
D. Drury	3	3	1	1	1	9	4	2	0	2	2	10	1	0	3	1	0	5																						1	.	.	1	1	1	
J. Eastwood	5	4	2	3	4	18													1	0	1	0	1	1	4	0	0	0	1	1	4	6								.	1	.	1	1	1	
H. Elliott	3	0	3	3	2	11	6	3	3	1	5	18														2	2	3	1	1	2	11								.	.	1	1	1	1	
B. Evans	6	2	1	0	0	9	4	0	2	1	1	8	5	5	0	2	1	13																						.	1	.	1	1	1	
J. Edwards	5	1	2	5	1	14													1	1	1	4	1	4	12								1	1	1	1	0	0	4	.	.	1	1	1	1	
D. Ferrier	7	2	4	3	0	16													1	4	1	1	4	2	13								1	0	2	0	0	0	3	.	.	1	1	1	1	
W. Fitzallen	6	5	4	1	0	16													1	1	1	1	1	3	8	1	1	1	0	0	0	3								.	1	.	1	1	1	
E. Fisher	5	3	2	0	2	12	5	4	4	0	0	13														0	0	1	4	0	0	5								.	1	.	1	1	1	
H. Fletcher	6	3	5	3	0	17													1	1	2	2	0	3	9	1	2	2	2	0	0	7								.	1	.	1	1	1	
T. Gardiner	7	3	1	0	3	14													2	0	0	1	0	0	3	1	0	0	1	1	0	3								.	1	.	1	1	1	
W. Gardiner	6	2	2	4	1	15													1	0	0	0	0	1	2	3	3	5	3	3	3	20								.	.	1	1	1	1	
W. Godsmark	5	6	7	0	2	20													2	1	2	0	1	2	8	4	1	2	0	1	2	10								.	.	1	1	1	1	
C. Hammerton	6	1	2	6	0	15													2	3	0	0	5	1	11								2	0	1	0	1	0	4	.	.	1	1	.	1	Prisoner.
T. Harriman	3	1	3	1	3	11	0	2	2	2	0	6	4	1	2	1	0	8																						1	.	.	1	1	1	
G. Harrison	10	1	3	3	3	20													.	.	.	.	.	.	.	.	.	.	.	.	.	.								1	.	.	1	1	1	In Prison.
J. Harrison	1	2	0	0	1	4	.	.	.	.	.	.	.	.	.	.	.	.																						1	.	.	1	1	1	
T. Hartwell	3	1	2	2	0	8	1	1	0	2	1	5	2	1	0	0	0	3																						1	.	.	1	1	1	
W. Hawksford	5	1	0	4	1	11	1	3	2	4	3	13														0	0	1	1	1	0	3								.	1	.	1	1	1	
H. Herbert	7	6	3	3	2	21													3	4	5	1	2	2	17								1	0	0	2	1	0	4	.	.	1	.	1	1	Hospital.
W. Hill	5	4	6	2	4	21	.												2	1	0	2	1	2	8	.	.	.	.	.	.	.								.	1	.	1	1	1	Absent.
W. Hills	6	5	6	5	3	25													1	1	2	2	0	0	6	1	0	0	1	0	0	2								.	1	.	1	1	1	
R. Howell	7	5	2	2	2	18													3	0	0	1	1	0	5	2	0	0	0	0	0	2								.	1	.	1	1	1	
T. Kellett	5	4	1	3	5	18													1	1	1	2	0	2	7	0	2	2	2	1	2	9								.	1	.	1	1	1	
J. Kelly	2	0	0	1	2	5	.	.	.	.	.	.	.	.	.	.	.	.																						1	.	.	1	1	1	Hospital.
T. Kemp	6	4	5	4	1	20													2	0	1	3	0	2	8	5	1	1	2	1	0	10								.	.	1	1	1	1	
R. Lawrence	8	7	4	2	1	22													3	1	1	0	0	3	8	1	3	0	0	2	2	8								.	1	.	1	1	1	
J. Laws	5	3	1	1	3	13	1	3	3	3	3	13														2	0	0	1	0	2	5								.	1	.	1	1	1	
W. Lennox	5	4	2	3	0	14													1	0	4	0	1	1	7	8	2	3	0	1	0	14								.	.	1	1	.	1	Prisoner.
W. Lloyd	3	3	5	3	1	15													1	2	1	4	3	3	14								0	3	0	1	1	0	5	.	.	1	1	1	1	
W. Lucas	5	4	7	6	2	24													1	4	2	0	0	0	7	5	2	2	0	1	1	11								.	.	1	1	1	1	
J. Lucas	5	5	4	4	3	21													2	0	2	4	2	4	14								0	0	0	0	0	0	0	.	.	1	1	1	1	
L. Lucas	2	4	1	3	0	10	4	3	5	1	1	14														2	0	0	1	1	1	5								.	1	.	1	1	1	
G. M'Guire	4	5	4	1	1	15													3	1	1	1	1	1	8	1	0	3	2	1	1	8								.	1	.	1	1	1	
W. M'Millan	9	6	2	4	2	23													1	1	2	0	3	4	11								3	2	0	1	0	0	6	.	.	1	1	1	1	
A. Manning	4	0	6	0	5	15													4	2	0	1	2	1	10								2	1	0	0	0	1	4	.	.	1	1	1	1	
J. May	5	5	6	2	0	18													0	1	1	1	0	1	4	0	0	0	1	1	0	2								.	1	.	1	1	1	
H. Martin	8	5	5	4	6	28													3	2	2	1	1	1	10								2	2	0	0	0	2	6	.	.	1	1	1	1	
W. Melbourne	5	3	3	3	1	15													.	.	.	.	.	.	.	4	0	2	2	3	2	13								.	.	1	1	1	1	Cells.
R. Minor	4	1	2	0	1	8	2	1	1	0	1	5	6	6	0	2	0	14																						.	1	.	1	1	1	
J. Mitchell	6	4	4	1	1	16													2	0	1	1	4	3	11								1	3	0	2	2	5	13	.	.	1	1	1	1	
J. Morgan	6	3	3	3	1	16													6	1	0	0	2	1	10								6	3	2	5	0	1	17	.	.	1	1	1	1	
J. Newton	.	.	.	.	.	.	1	1	3	1	3	9	5	0	2	2	1	10																						1	.	.	.	1	1	On Pass.
T. Owens	.	.	.	.	.	.	0	0	0	3	0	3	1	1	1	2	0	5																						1	.	.	.	1	1	Prisoner.
D. Palmer	7	2	5	3	1	18													0	0	2	0	0	1	3	0	1	2	1	2	0	6								.	1	.	1	1	1	
T. Phillips	5	3	3	3	2	16													0	1	0	1	0	4	6	1	2	5	0	1	1	10								.	.	1	1	1	1	
W. Plumb	2	2	3	2	2	11	3	2	1	1	1	8	3	2	2	1	3	11																						1	.	.	1	1	1	
W. Poleman	3	5	1	5	2	16													4	1	3	0	1	1	10								2	1	1	0	0	3	7	.	.	1	1	1	1	
W. Poynter	.	.	.	.	.	.	2	3	1	1	1	8	0	0	1	0	0	1																						1	.	.	1	1	1	Absent.
J. Poynter	4	3	3	3	1	15													.	.	.	.	.	.	.	2	0	0	1	0	2	5								.	1	.	1	1	1	Prisoner.
B. Pulley	2	4	1	1	1	9	1	3	3	3	3	14														1	3	4	0	1	0	9								.	1	.	1	1	1	
R. Rackstraw	6	4	4	4	3	21													1	0	2	2	1	2	8	1	0	1	0	0	1	3								.	1	.	1	1	1	
D. Robertson	2	4	1	1	3	11	1	3	5	2	2	13														3	3	3	1	0	0	10								.	.	1	1	1	1	
J. Rutter	8	0	3	1	1	13	4	3	1	3	1	12	3	1	0	1	0	8																						1	.	.	1	1	1	
W. Snape	5	2	2	1	2	12	5	4	1	3	3	15														0	2	1	0	1	1	5								.	1	.	1	1	1	
H. Snart	5	5	6	2	0	18													0	1	1	2	0	0	4	.	.	.	.	.	.	.								.	1	.	1	1	1	Hospital.
H. Smith	4	2	3	0	1	10	0	2	1	1	2	6	1	4	2	1	1	9																						1	.	.	1	1	1	
J. Short	3	5	1	5	3	17													2	5	1	3	0	0	11								3	3	2	0	0	1	9	.	.	1	1	1	1	
T. Smith	2	1	0	2	0	5	5	2	3	0	0	10	4	3	3	0	1	11																						1	.	.	1	1	1	
J. Stevens	.	.	.	.	.	.	0	0	0	1	0	1	1	2	1	2	5	11																						1	.	.	1	1	1	Hospital.
J. Tapsill	9	2	2	3	0	16													1	1	0	0	1	1	4	3	2	0	2	0	0	7								.	1	.	1	1	.	Absent.
Jos. Taylor	3	5	4	2	0	14													1	0	1	1	1	5	9	2	1	1	0	1	0	5								.	1	.	1	1	1	
W. Taylor	7	4	3	2	1	17													0	1	0	0	1	4	6	2	1	1	0	1	0	5								.	1	.	1	1	1	
J. Talbot	8	6	7	1	2	24													1	4	2	2	1	2	12								5	2	3	3	3	1	17	.	.	1	1	1	1	
W. Trott	2	2	1	0	0	5	4	1	3	1	1	10	1	3	1	0	2	7																						1	.	.	1	1	1	
T. Trowbridge	2	6	2	1	2	13													.	.	.	.	.	.	.	1	0	0	0	0	1	2								.	1	.	1	1	1	Cells.
T. Tyler	2	1	2	0	4	9	0	4	1	2	3	10	3	1	1	0	3	8																						1	.	.	1	1	1	
S. Vizney	4	3	4	3	0	14													0	0	1	1	1	3	6	4	3	0	2	2	2	13								.	.	1	1	1	1	
F. Ward	3	5	2	4	1	15													3	1	0	1	1	2	8	1	1	3	0	0	1	6								.	1	.	1	1	1	
W. Wilcock	2	1	0	3	1	7	2	1	0	2	1	6	2	3	6	5	0	16																						.	1	.	1	1	1	
R. Whitlock	3	7	1	2	1	14													4	1	0	1	1	3	10								4	0	2	0	0	1	7	.	.	1	1	1	1	
R. Willgress	0	5	1	3	0	10													2	3	0	0	1	3	9	1	1	2	1	3	1	9								.	1	.	1	1	1	
W. Wilson	7	5	1	1	0	14													1	2	3	2	2	2	12								1	1	1	0	0	3	6	.	.	1	1	1	1	
Totals	431	300	236	217	167	1400	106	79	89	75	32	412	71	83	46	13	87	295	91	81	68	66	64	117	487	76	66	66	54	43	81	366	43	23	30	16	11	17	186	27	47	33	106	106	106	

	3rd Class, 1st Period	3rd Class, 2nd Period	3rd Class, 3rd Period	2nd Class, 2nd Period	2nd Class, 3rd Period	1st Class, 3rd Period
[Signature o]f Captain or Officer commanding	T. Berry, Captain.	T. Berry, Captain.	T. Berry, Captain.	T. Berry, Captain.	T. Berry, Captain.	T. Berry, Captain.
[Signature o]f Officer Instructor	C. Lucas, Officer-Instructor.	C. Lucas, Officer-Instructor.	C. Lucas, Officer-Instructor.	C. Lucas, Officer-Instructor.	C. Lucas, Officer-Instructor.	C. Lucas, Official Instructor.

RECAPITULATION.	Men.	Rounds.	Points.	Average.	Signature of Captain.	Signature of Officer-Instructor.
[Obta]ined by the Company in the Third Class	107	20	1400	13·08	T. Berry, Captain.	C. Lucas, Officer-Instructor.
ditto in File-firing and Volleys	102	10	757	7·48	T. Berry, Captain.	C. Lucas, Officer-Instructor.
ditto in Skirmishing, 1 Practice	103	10	341	3·28	T. Berry, Captain.	C. Lucas, Officer-Instructor.
ditto in Skirmishing, 2 Practice	103	10	437	4·25		
Totals	..	50	..	*28·09	*This figure denotes the merit of the Company.	

T. Berry, Captain.

C. Lucas, Captain, Officer-Instructor.

Form 93.   No. 10 Company, 1st Battalion, 91st Regiment.   Company, Judging Distance R

JUDGING DISTANCE PRACTICE RETURN of Captain J. LUCAS' Company for the Year 1855. Date July 10th 1855.

Number	Rank and Names		3rd Class																		2nd Class										1st Class					Classification at Conclusion of 3rd Period			REMARKS
			1st Period. Practice of Company in 3rd Class.					2nd Period. Practices.					3rd Period. Practices.					1st Period. Practices.					2nd Period. Practices.					3rd Period. Practices.								
			1 Answers. 4 Points.	2 4	3 4	4 4	Total Points.	1 4	2 4	3 4	4 4	Total Points.	1 4	2 4	3 4	4 4	Total Points.	1 4	2 4	3 4	4 4	Total Points.	1 4	2 4	3 4	4 4	Total Points.	1 4	2 4	3 4	4 4	Total Points.	3rd Class	2nd Class	1st Class	
1	Serjeant	T. Andrews	6	3	8	10	29										3	3	4	2	16						1	5	2	8	14	..	..	1	
2	,,	J. Jones	1	3	3	4	10	3	4	4	3	16											4	1	2	4	12						..	1	..	
3	,,	A. King	4	1	5	4	10	3	2	1	7	13	4	5	3	4	16																..	..	1	
4	,,	J. Lambkin	..	..	..	..	..	..	..	..	..	..	3	4	3	0	10																..	1	..	In Hospital 1st and 2d Periods.
5	Corporal	M. Manning	5	4	5	3	17											..	..	..	..	..	3	4	2	5	14						..	1	..	Absent, 3d Period.
6	,,	W. Nankin	4	5	5	4	19										3	4	5	6	15											..	1	..	In Hospital, 3d Period.
7	,,	C. Orr																																		Prisoner.
8	,,	J. Palmer	4	7	5	10	26						2	1	6	6	15	2	7	1	2	12						13	5	4	6	27	..	..	1	
9	Private	J. Peterson	3	0	4	0	7	2	8	4	5	17						3	4	1	2	10	3	5	5	3	16						..	..	1	
10	,,	J. Reiter	2	1	6	8	13	3	8	4	5	17						1	7	7	3	18						2	3	0	6	15	..	..	1	
11	,,	J. Snart	11	10	9	11	41											0	8	6	14		1	3	5	5	13						..	..	1	
12	,,	J. Snape	2	11	6	8	34																										1	..	..	
13	,,	J. Stevens	3	4	3	0	11	3	2	1	3	11	0	2	1	6	9																..	1	..	
14	,,	A. Taylor	1	3	5	4	10	3	4	1	7	15	4	1	3	7	15																..	..	1	
15	,,	J. Tyler	2	10	9	11	20											10	6	3	0	19						4	8	3	5	20	..	1	..	
16	,,	T. Vicary	2	1	4	6	10	10	1	8	6	35											3	3	3	4	14						..	1	..	
17	,,	A. Ward	2	10	7	8	34	..	..	..	..	..						5	2	7	6	20						3	3	2	6	22	1	..	1	
18	,,	T. Walton	..	..	..	..	..	4	3	2	1	10																					1	..	..	Prisoner, 1st Period; Hospital, 3d Period

| Total | | | | | | | | | | | | | | | | | | | | | | | | | | | | | | | | | | | | | |

Signature of Captain or Officer commanding Company

Signature of Officer-Instructor

Captain

Progress Return of the Instruction in Musketry in the above Battalion from 1st to 30th June 1855. Templemore, 1st July 1855.

Strength of each Company or Depôt of Depôt Battalion, not including Recruits.			Companies of Regiments, or Depôts of Depôt Battalion.	Number under instruction, as also the Number who have concluded any of the Practices.			Stages of Instruction through which the Men have been exercised during the above Period.																Remarks.
							Preliminary Drills.						Target Practice.									Judging Distance Practice. Concluded or in Progress.		
Officers.	Serjeants.	Rank and File.		Officers.	Serjeants.	Rank and File.	Theoretical Principles.	Cleaning of Arms.	Target Drill.			Judging Distance Drill.	Average Points per Man in each Class in the following Practices.						File firing and Volley.	Skirmishing.		1st Period.	2nd Period.	3rd Period.
									Aiming Drill.	Position Drill.			1st Period. 3rd Class.	2nd Period.		3rd Period.				1st Practice.	2nd Practice.			
														3rd Class.	2nd Class.	3rd Class.	2nd Class.	1st Class.						
3	5	80	No. 1	1	1	21	Concluded	Concluded	Concluded	Concluded		Concluded	10·16	11·19	8·17	..	..	..	..	..	..	Concluded	Concluded	..
3	5	70	No. 2	1	1	21	In progress	In progress	In progress	In progress		In progress	9·72	10·15	8·19	..	..	..	..	..	..	Concluded	Concluded	..
3	5	86	No. 3	1	1	20	Inprogress	In progress	In progress	In progress		In progress	10·89	9·87	9·57	..	..	..	..	..	..	Concluded	Concluded	..
3	5	80	No. 4	1	1	18	Concluded	Concluded	Concluded	Concluded		Concluded	10·89	9·87	9·57	..	..	..	..	..	..	Concluded	Concluded	..
3	5	80	No. 5	1	1	20	In progress	In progress	In progress	In progress		In progress	11·18	10·42	9·81	..	..	..	..	..	..	Concluded	Concluded	..
3	4	79	No. 6	1	1	24	In progress	In progress	In progress	In progress		In progress	8·74	10·60	8·21	..	..	..	..	..	..	Concluded	Concluded	..
3	5	80	No. 7	1	1	20	In progress	In progress	In progress	In progress		In progress	9·47	In progress	..	..	..	..	..	..	..	Concluded	In progress	..
3	4	82	No. 8	1	1	17	In progress	In progress	In progress	In progress		Concluded	10·98	In progress	..	..	..	..	..	..	..	Concluded	In progress	..
3	5	79	No. 9	1	1	20	In progress	In progress	In progress	In progress		Concluded	11·63	In progress	..	..	..	..	..	..	..	Concluded	In progress	..
3	4	84	No. 9	1	1	21	In progress	In progress	In progress	In progress		Concluded	7·48	In progress	..	..	..	..	..	..	..	Concluded	In progress	..
3	4	76	No.10	1	1	19	Concluded	Concluded	Concluded	Concluded		Concluded	8·96	In progress	..	..	..	..	..	..	..	Concluded	In progress	..
30	46	796	Totals	15	15	300																		
			Recruits { 1	..	..	20	Concluded	Concluded	Concluded	Concluded		Concluded	In progress	..	..	..	..	..	..	..	..			
			2	..	..	40	In progress	In progress	In progress	In progress		In progress	..	..	..	..	..	..	..	..	..			
			3																					
			4																					

Number of Men in each Class who have concluded the Periods here named.

	Target Practice.			Judging Distance Practice.			Remarks.
	3rd Class.	2nd Class.	1st Class.	3rd Class.	2nd Class.	1st Class.	
1st Period	100	109	..	92	117	..	
2nd Period	30	64	12	10	59	40	
3rd Period	..	..	..	..	..	..	

C. Lucas, Captain, Instructor.
J. Lawrow, Lieut.-Col., Commanding 97th Regiment.

PLATE 16.

_____ BATTALION, _____ REGIMENT.

Annual Target Practice Return of the above Corps for the Year 18___.

Date _____

Companies.	PRACTICE BY COMPANIES.												Total average points per man obtained in the foregoing practices, (Nos. 1 to 4.)	PRACTICE BY CLASSES.												No. of men in each Class at the termination of practice, exclusive of recruits				Total strength including N.C. Officers but exclusive of recruits.	Number of Recruits.	REMARKS, Showing why the Men who are returned "Not exercised" did not practise.
	As a Company in 3rd Class, each man firing 20 rounds. 1			File and volley firing, 10 rounds per man. 2			1st practice, 10 rounds per man. 3			Skirmishing. 2nd practice, 10 rounds per man. 4				2nd Period, each man firing 20 rounds.						3rd Period, each man firing 20 rounds.							Third.	Second.	First.	Not exercised.			
	Men.	Points.	Average.	Men.	Points.	Average.	Men.	Points.	Average.	Men.	Points.	Average.		3rd Class. Men.	Points.	Average.	2nd Class. Men.	Points.	Average.	3rd Class. Men.	Points.	Average.	2nd Class. Men.	Points.	Average.	1st Class. Men.	Points.	Average.						
No. 1	90	1740	19·33	90	1000	11·11	88	374	4·25	88	363	4·12	38·81	110	212	21·20	80	660	8·25	5	71	14·20	40	3659	9·01	45	2906	44	2	29	59	3	93	11	
2	96	1608	16·75	91	975	10·71	90	360	4·00	90	495	5·33	36·79	6	91	15·16	90	690	7·66	3	40	13·33	63	5158·17		30	1966·53	2	34	60	1	97	8	
3	81	1224	15·11	80	810	10·12	81	423	5·22	81	297	3·77	34·22	8	138	17·25	72	656	9·25	4	51	12·75	55	4909·27		22	1577·13	3	41	37	4	85	16	
4	100	1720	17·20	96	960	10·00	95	456	4·80	95	285	3·00	35·00	9	174	18·22	88	748	8·50	6	74	12·33	64	5448·50		30	1866·20	5	57	38	··	100	4	
5*	··	··	··	··	··	··	··	··	··	··	··	··	··	··	··	··	··	··	··	··	··	··	··	··	··	··	··	··	··	··	··	··	··	··	On Detachment.
6*	··	··	··	··	··	··	··	··	··	··	··	··	··	··	··	··	··	··	··	··	··	··	··	··	··	··	··	··	··	··	··	··	··	··	
7*	··	··	··	··	··	··	··	··	··	··	··	··	··	··	··	··	··	··	··	··	··	··	··	··	··	··	··	··	··	··	··	··	··	··	
8*	··	··	··	··	··	··	··	··	··	··	··	··	··	··	··	··	··	··	··	··	··	··	··	··	··	··	··	··	··	··	··	··	··	··	
9*	··	··	··	··	··	··	··	··	··	··	··	··	··	··	··	··	··	··	··	··	··	··	··	··	··	··	··	··	··	··	··	··	··	··	
10	103	1409	13·68	102	757	7·42	103	541	5·25	103	437	4·24	30·59	42	412	7·80	57	483·8	8·47	30	264	8·80	50	3507·00		30	1286·80	27	47	33	··	107	··	
Total	470	7701		459	4502		457	2154		457	1877			75	1027		387	3937		48	500		272	22264		147	957	39	208	227	8	482	39	
Total points divided by No. of men.			16·38			9·85			4·71			4·10	35·04			13·69			8·36			10·41			8·32			6·51							

Average points per man of the battalion in the First period, 3rd class . . . . †16·38

    "   "   "   in File and Volley firing . . . . 9·85

    "   "   "   in Skirmishing { 1st Practice . . . 4·71

                                          { 2nd   "   . . . 4·10

                                           Total . . . . 35·04

No. of rounds per man fired to complete the annual course of practice. Rounds { _____ }

Extent of Range _____ yards.

Practice { Commenced _____ 18

             { Concluded _____ 18

_____ Commanding Officer.

_____ Officer-Instructor.

_____ Inspecting Officer.

PLATE 17.

K. _____ BATTALION, _____ REGIMENT.

Annual Judging Distance Practice Return of the above Corps for the Year 18___. Date _____

COMPANIES.	1ST PERIOD. Practice as a Company in 3rd Class.			2ND PERIOD. 3rd Class.			2nd Class.			3rd Class.			3RD PERIOD. 2nd Class.			1st Class.			Number of Men in each Class at conclusion of the Practice.				Total strength, including Non-commissioned Officers.	REMARKS, Showing the causes why the Men who are returned "Not exercised" did not practise.
	Men.	Points	Average.	Men.	Points.	Average.	Men.	Points.	Average.	Men.	Points	Average.	Men.	Points.	Average.	Men.	Points.	Average.	1st.	2nd.	3rd.	Not exercised.		
No. 1	90	1877	20·86	14	241	17·21	75	1388	18·50	11	189	17·18	56	1036	18·50	23	353	15·34	6	44	40	3	93	
No. 2	96	1984	20·66	17	281	16·52	79	1548	19·59	10	167	16·70	65	1267	19·48	20	331	16·55	3	51	42	1	97	
No. 3	81	1586	19·58	13	203	15·61	66	1351	20·46	9	142	15·77	42	852	20·28	30	529	17·63	4	28	49	4	85	
No. 4	100	1971	19·71	19	291	15·31	81	1445	17·83	13	207	15·92	60	1296	21·60	25	467	18·68	3	36	62	··	100	
No. 5	··	··	··	··	··	··	··	··	··	··	··	··	··	··	··	··	··	··	··	··	··	··	··	
No. 6	··	··	··	··	··	··	··	··	··	··	··	··	··	··	··	··	··	··	··	··	··	··	··	
No. 7	··	··	··	··	··	··	··	··	··	··	··	··	··	··	··	··	··	··	··	··	··	··	··	
No. 8	··	··	··	··	··	··	··	··	··	··	··	··	··	··	··	··	··	··	··	··	··	··	··	
No. 9	··	··	··	··	··	··	··	··	··	··	··	··	··	··	··	··	··	··	··	··	··	··	··	
No. 10	··	··	··	··	··	··	··	··	··	··	··	··	··	··	··	··	··	··	··	··	··	··	··	
Total -	367	7418		63	1016		301	5732		43	705		223	4451		98	1680		15	159	193	8	375	
Total points divided by No. of Men.			20·21			16·12			19·04			16·39			19·95			17·14						

_____ Commanding Officer.

_____ Officer-Instructor.

_____ Inspecting Officer.

D 14

www.ingramcontent.com/pod-product-compliance
Lightning Source LLC
LaVergne TN
LVHW091318080426
835510LV00007B/540